Long Term Loan to
HE Persinger
Please Return

COMPREHENSIVE BIOCHEMISTRY

ELSEVIER PUBLISHING COMPANY

335 Jan van Galenstraat, P.O. Box 211, Amsterdam

AMERICAN ELSEVIER PUBLISHING COMPANY, INC.

52, Vanderbilt Avenue, New York, N.Y. 10017

Library of Congress Catalog Card Number 62–10359

ISBN 0–444–41024–4

With 63 plates, 4 figures and 4 tables

PRINTED IN THE NETHERLANDS

COMPREHENSIVE BIOCHEMISTRY

COMPREHENSIVE BIOCHEMISTRY

SECTION I (VOLUMES 1–4)
PHYSICO-CHEMICAL AND ORGANIC ASPECTS OF BIOCHEMISTRY

SECTION II (VOLUMES 5–11)
CHEMISTRY OF BIOLOGICAL COMPOUNDS

SECTION III (VOLUMES 12–16)
BIOCHEMICAL REACTION MECHANISMS

SECTION IV (VOLUMES 17–21)
METABOLISM

SECTION V (VOLUMES 22–29)
CHEMICAL BIOLOGY

SECTION VI (VOLUMES 30–33)
A HISTORY OF BIOCHEMISTRY

(VOLUME 34)
GENERAL INDEX

COMPREHENSIVE BIOCHEMISTRY

EDITED BY

MARCEL FLORKIN

Professor of Biochemistry, University of Liège (Belgium)

AND

ELMER H. STOTZ

Professor of Biochemistry, University of Rochester, School of Medicine and Dentistry, Rochester, N.Y. (U.S.A.)

VOLUME 30

A HISTORY OF BIOCHEMISTRY

Part I. Proto-Biochemistry

Part II. From Proto-Biochemistry to Biochemistry

by

MARCEL FLORKIN

ELSEVIER PUBLISHING COMPANY

AMSTERDAM · LONDON · NEW YORK

1972

GENERAL PREFACE

The Editors are keenly aware that the literature of Biochemistry is already very large, in fact so widespread that it is increasingly difficult to assemble the most pertinent material in a given area. Beyond the ordinary textbook the subject matter of the rapidly expanding knowledge of biochemistry is spread among innumerable journals, monographs, and series of reviews. The Editors believe that there is a real place for an advanced treatise in biochemistry which assembles the principal areas of the subject in a single set of books.

It would be ideal if an individual or small group of biochemists could produce such an advanced treatise, and within the time to keep reasonably abreast of rapid advances, but this is at least difficult if not impossible. Instead, the Editors with the advice of the Advisory Board, have assembled what they consider the best possible sequence of chapters written by competent authors; they must take the responsibility for inevitable gaps of subject matter and duplication which may result from this procedure.

Most evident to the modern biochemist, apart from the body of knowledge of the chemistry and metabolism of biological substances, is the extent to which he must draw from recent concepts of physical and organic chemistry, and in turn project into the vast field of biology. Thus in the organization of Comprehensive Biochemistry, sections II, III and IV, Chemistry of Biological Comjounds, Biochemical Reaction Mechanisms, and Metabolism may be considered classical biochemistry, while the first and fifth sections provide selected material on the origins and projections of the subject.

It is hoped that sub-division of the sections into bound volumes will not only be convenient, but will find favour among students concerned with specialized areas, and will permit easier future revisions of the individual volumes. Toward the latter end particularly, the Editors will welcome all comments in their effort to produce a useful and efficient source of biochemical knowledge.

M. Florkin
Liège/Rochester E. H. Stotz

PREFACE TO SECTION VI

(Volumes 30–33)

In the many chapters of previous sections of *Comprehensive Biochemistry* covering organic and physicochemical concepts (Section I), chemistry of the major constituents of living material (Section II), enzymology (Section III), metabolism (Section IV), and the molecular basis of biological concepts (Section V), authors have been necessarily restricted to the more recent developments of their topics. Any historical aspects were confined to recognition of events required for interpretation of the present status of their subjects. These latest developments are only insertions in a science which has had a prolonged history of development.

Section VI is intended to retrace the long process of evolution of the science of Biochemistry, framed in a conceptual background and in a manner not recorded in recent treatises. Part I of this section deals with Proto-biochemistry or with the discourses imagined concerning matter-of-life and forces-of-life before molecular aspects of life could be investigated. Part II concerns the transition between Proto-biochemistry and Biochemistry and retraces its main landmarks. In Part III the history of the identification of the sources of free energy in organisms is depicted, and Part IV is devoted to backgrounds in the unravelling of biosynthetic pathways. While these latter parts are concerned with the molecular level of integration, Part V is more specifically directed toward the history of molecular interpretations of physiological and biological concepts, and of the origins of the concept of life as the expression of a molecular order.

The *History* narrated in Section VI thus leads to the thresholds of the individual histories in the recent developments recorded by the authors of Sections I–V of *Comprehensive Biochemistry*.

Liège/Rochester

M. Florkin
E. H. Stotz

CONTENTS

VOLUME 30

A HISTORY OF BIOCHEMISTRY

Chapter 3. Respiratory Theory of "Vital Heat" and Phlogistonic Proto-Biochemistry

Chapter 4. "Dynamic Permanence" and "Assimilation" Prior to the Chemical Revolution, p. 97

Part II
From Proto-Biochemistry to Biochemistry

Chapter 5. Metabolic Theories of Lavoisier and his Followers

Chapter 6. The Nature of Alcoholic Fermentation, the "Theory of the Cells" and the Concept of the Cells as Units of Metabolism

Chapter 7. The Rise and Fall of Liebig's Metabolic Theories

Chapter 8. The Intracellular Location of Metabolic Changes

Chapter 9. The Reaction Against "Analysm": Antichemicalists and Physiological Chemists of the 19th Century

Chapter 10. Dynamic Permanence after Liebig, and the End of the Myth of Direct Assimilation

Chapter 11. From Forces-of-Life to Bioenergetics

Chapter 12. Life Banned from Organic Chemistry

Chapter 13. Biocatalysis and the Enzymatic Theory of Metabolism

Chapter 14. The Dark Age of Biocolloidology, p. 279

Chapter 15. Recognition of the Proteins as Truly Defined Macromolecules, p. 285

Chapter 16. Biochemists Find Their Way into the Cell

Appendices

Section VI

A HISTORY OF BIOCHEMISTRY

Vol. 30. A History of Biochemistry
Part I. Proto-Biochemistry
Part II. From proto-biochemistry to biochemistry

Vol. 31. A History of Biochemistry
Part III. History of the identification of the sources of free energy in organisms

Vol. 32. A History of Biochemistry
Part IV. The unravelling of biosynthetic pathways

Vol. 33. A History of Biochemistry
Part V. History of molecular interpretations of physiological and biological concepts, and of the origins of the conception of life as the expression of a molecular order

List of Illustrations

Plates

(Unless otherwise stated, the portraits belong to the author's own personal collection.)

Figures

General Introduction

Biochemistry (a translation of the term "Biochemie", coined by Hoppe-Seyler[1] in 1877) "covers all molecular approaches to biology"[2] and its history cannot be entirely separated from the history of biology itself. There was no real science of biochemistry before the chemists could study the nature, structure and properties of the molecular and macromolecular components of organisms, and before experimentation could be interpreted with the help of the paradigms* of chemical and physical sciences. It is only on this basis that a scientific discipline could become concerned with molecular transformations and movements taking place in organisms, and with the molecular basis of energy supply for the activity of cells, as well as for the plastic phenomena responsible for the transfer of information controlling self-replication assembly and the existence of the molecular signals subserving the sequence of the interactions of molecules, of cells and of organisms. Biochemistry has been for some time, and sometimes still is, despised by chemists, as concerned with "slime", and it was long considered as a mere handmaiden by physiologists. It slowly succeeded in acquiring an independent status as an autonomous science, characterized by its specific scope and to which a large number of scientists are devoted (see Appendix 1, p. 321). Many specialized periodicals are now recording the archives of its accomplishments. As the mass of data that is continuously acquired is increasing at a fast rate, as attested by the size of this *Comprehensive Biochemistry*, it has appeared to its editors that the time was ripe for a historical treatment of the progress of biochemistry. It is to the sequence of discoveries that have led to the biochemical theories of our time, to the progressive "molecularization" of biology, and to the influence on these biochemical theories, of paradigm shift in cosmology, biology, physics and chemistry, that this study is essentially devoted.

Since the formulation of the cell theory in the red-letter year 1839, the promised land of biochemistry has become the understanding of the structure, function and interactions of the biochemical systems of the cells and of the organisms resulting from this association and integration, *i.e.* those

* Paradigms are defined by Kuhn[3] as "universally recognizable scientific achievements that for a time provide model problems and solutions to a community of practitioners".

References p. 17

aspects which Lehninger[4] refers to as "the molecular logic of the living state" and Jacob[5] as "la logique du vivant".

One of the duties of the historian of a science is to narrate. He must also extricate from the tangled skein of the chronological sequence of theories the threads connected to a knowledge more adequate than was previous knowledge, and recognize the origins of these threads. As well as the origins of ratified knowledge, he must recognize the virtues of obsolete knowledge and obsolete theories and the duration of their heuristic value, if they had such a value, and identify the process of their destruction. Occasionally scientific theories are in relation with the impact of external, social or economic factors, which must be recognized if of importance. In any case, the internal process of maturation of a theory, commanding its insertion into the conceptual system of science, has to be retraced: to recognize the genetic epistemology of the concepts concerned is one of the duties of the historian. He therefore differs from the scientist who deals with the object of his discourse and who, when relating the last steps of the development of the subject of his study, deals with a short period of continuously related recent approaches. But the science to which the object belongs has become a science through a long process of a succession of hypotheses and verifications which is forgotten (and does not appear any more in the treatises recording the last situation of the conceptual system of the science concerned*) when the results of this long process are by common agreement inserted into this system. As noted by Canguilhem, generally speaking science is the study of an object that has no history, that is no history, whereas the history of the science is a history of the discourses held on the nature of this object, and these discourses did not at once provide a knowledge adequate to the object. To quote Canguilhem:

> "L'objet du discours historique est, en effet, l'historicité du discours scientifique, en tant que cette historicité représente l'effectuation d'un projet intérieurement normé, mais traversée d'accidents, retardée ou détournée par des obstacles, interrompue de crises, c'est-à-dire de moments de jugement et de vérité". (ref. 6)

The project which has resulted in the development of the molecular approaches to organisms cannot, as said above, be separated from the general concern with these organisms, except by an emphasis put on matter-of-life

* This does not apply to the admirable synthesis of biochemistry written by Lehninger[4], in which, besides his competence as a pioneer in a number of fields of biochemistry, he displays his literary gifts as well as a fine sense of philosophy and history.

and on forces-of-life. A number of themes return time and again, since antiquity, in the approach to living organisms by the human mind, a realistic approach of a dominant visual character at the start. The character of this approach is expressed in such classical queries as:—What does it look like? How is it made inside? Of what is it made? How does it work? leading to the concern with shape, structure, composition and functioning, and to scientific disciplines concerned with different levels of magnitude*, such as systematics, anatomy, histology, cytology, physiology, biochemistry, etc. The same classical queries obtain again for each level of magnitude. The question—How does it work? for instance, may lead to *anatomia animata*, experimental physiology, physiological histology, cell physiology, molecular biology, etc. The fascination of the human mind with the third dimension finds its expression in all fields of intellectual activity. It is at the origin of architecture and of sculpture. In biology, this thematic aspect is first chronologically expressed in the description of the external surfaces of whole organisms.

Structure, another aspect of the concern with a tridimensional consideration, naturally follows the consideration of shape. Classical sculpture deals almost exclusively with shape delimited by external surfaces, but it is to be noted that, in contradiction with classical tradition, modern sculpture has opposed "open sculpture" to what has been called "sculpture for the blind", and has given pre-eminence to the relation of structure with space. The reaction against monolithic shape has taken another, more subtle, form with the internal structure introduced by the sculptor Berrocal, who composes an aesthetic shape by an assembly of parts, each one of aesthetic quality. After having put the parts together, the owner of the sculpture keeps a knowledge of the internal structure and escapes the dullness of the shapeless internal material and the resulting morosity. The same kind of attitude of the mind is found in Buffon's work (see Chapter 4). He refers, in the consideration of an organism, to an external mould similar to the cast of the sculptor (shape) but feels the necessity of an "interior mould" (structure). Shape, structure and composition mark the steps of the reduction process in the human mind in its approach to macroscopical objects. The primitive man, for instance, when attracted by the external shape of a fruit, may crush it and find a pulp and a stone as elements of structure.

* The concept of "integrative level" derives from the same tendency of the human mind. It is found in a number of philosophical systems: dialectal materialism of Marx and Engels, organic mechanism of Whitehead, holism of Smuts, etc.

References p. 17

When tasting the fruit, he may find it sweet or bitter, a first approach to composition. But it should be against historical reality to accept that the consideration of organisms has always moved along a path of reduction dealing sequentially with shape, structure and composition of parts of progressively smaller size. If, at the scale of whole organisms, the trail of research has successively moved from shape to structure and to composition, on the contrary in the consideration of macromolecular units composing the organism, the trail of thought has moved from composition to structure and orderliness and finally to shape, the difference of sequence deriving from the different availabilities of concepts and of techniques of investigation.

The interest in shape, combined with the common trivial knowledge of a flux of matter through the organism, gives rise to the concept of "dynamic permanence", as an adult organism maintains its shape in spite of continual ingestion of food and elimination of excreta. This theme becomes analyzed in two phases, a constructive aspect and a destructive aspect (composition and decomposition). To relate the knowledge of the consumption of food and the constructive aspect of dynamic permanence, the human mind has imagined the long-lasting theme of direct assimilation, according to which the constituents of foods are, either as such or after slight modification, inserted into matter-of-life, replacing its worn-out parts. The composition of organisms, one of the continually recurring themes, is studied in relation to the contemporary stage of the knowledge of matter. The organisms were first considered as formed, as was the whole of the universe, of associations of the four elements. After the chemical revolution, the organisms were considered as composed of chemical elements. The success of the concept of species in animal and plant taxonomy, at the level of external shape, prompted, through the work of Lavoisier and of Chevreul, a search for chemical species, the "proximate principles". When the concepts of molecules and of macromolecules were defined, the composition of the organisms was expressed in terms of their units.

Another ever-recurring theme is the analogy of the organism with a machine. This theme implies that neither machine nor organism

"could be expected to arise spontaneously from a mixture of their component molecules; that is without some special conditions or events extraneous to the simple chemical mixture itself." (ref. 7)

The discovery of a number of self-assembling systems in organisms has rendered this analogy untenable.

Another general theme is the comparison of the organism with a flame. The vital flame of the ancient philosophers was situated in the heart, where it maintained the innate heat. From the heart the heat as well as the vital spirits were carried to all parts. The breath, *pneuma*, first considered as cooling the heart's heat, was later considered as feeding the flame to which it was brought by the anastomoses between the pulmonary vein and the bronchia, the latter carrying away the products of the combustion. As noted by Belloni,

"In the apse mosaïc of "St. Apollinare in Classe" basilica at Ravenna (Italy), dating back to the VIIth century A.D. and dedicated to the transfiguration, the Almighty's hand and voice are symbolyzed[9]: a heart in direct communication with the laryngotracheal tube is the symbol of the voice: it is from the heart that, together with the inspired air, proceeds the voice with the weight of feelings it includes; the warmth of these feelings is associated with the physical heat . . ." (ref. 8)

The flame of the heart is still kept in the ex-voto of our days. It was to shift to the lungs and after that to the blood and finally to all tissues. As the organisms reproduce their kind, they must transfer to their offspring a principle that leads and controls the development of the young organism. This principle has been successively formulated as soul, as vital force and now as the information carried by the genes.

The themes that have recurred so often in the biochemical approaches to organisms since ancient Greek proto-biochemistry belong to the structure of this approach and to the normative factors of its development. But it is wrong to overvalorize the heuristic virtue of these themes. Much more important were the surprises, the ruptures that arose along the accomplishment of the project and oriented the trail of ideas and methodologies in unexpected directions. Another characteristic of the human mind, the quest for universality, was gratified by one of these ruptures, the introduction of the concept of the cell, as the universal unity of metabolism as well as of form, and as a basis of development. At the molecular level, food was recognized, not as directly replacing worn-out parts as had been imagined, but as sharing with solar light the quality of source of the flux of energy going through the organisms. The concept of biocatalysis came as another unexpected rupture, and the enzymatic theory of metabolism gave the rationale of composition and decomposition. Enzymes themselves could then become subject to studies of composition, structure and shape after they were recognized to be of the nature of proteins, themselves identified as definite chemical compounds. The quest for universality was

References p. 17

again satisfied when the flux of energy through organisms was situated in the general energetics of the universe.

First considered as moved by innate heat and later by the heat generated by respiratory combustions, the organisms have been recognized as isothermal chemical engines, and thermodynamically as open systems, the go of which is derived from the high-energy bonds of some phosphorus derivatives.

To the flux of matter, and to the flux of energy, was added a flux of information, implemented by the accomplishment of a coded programme of protein synthesis in which the macromolecule of DNA ensures the plastic genesis of molecules and the large amount of information they carry and which rules their self-assembly and their interrelations. From the whole development of biochemical knowledge by these ruptures accomplished by recourse to more and more penetrating techniques, transcending the constant themes of the human mind, has emerged a form of highly documented reduction, in opposition to the naive reductionisms of Paracelsus, of Van Helmont, of Descartes or of Schwann. In the perspective of this reduction the transformation of energy in organisms at the level of the cells is shown to take place at the molecular level, in a very complex net of molecular events resulting in the generation of free energy sources. Here the sequence of actions subserving the transformations of energy at the molecular level is not imagined but it is clarified in a critical way. Not only has the flux of energy of which an organism is the seat been recognized as actually taking place at the molecular level, but the biological concepts of heredity, of development and of evolution and the physiological concepts of the functions of the organisms have also been recognized as based on molecular interactions. As Edsall writes:

"Since about 1950 biology has assumed a central role among the sciences, and biochemistry has taken a similar central place within biology, from genetics to taxonomy." (ref. 10)

Our theory of life is a molecular theory. It may be defined historically as an outcome of a progressive "molecularization" of life. As we have defined it, biochemistry, covering all molecular aspects of biology, becomes more and more the science of life. It takes advantage of all kinds of methodology, not only the chemical ones, but in spite of this it deserves more than ever the name of biochemistry: as chemistry is the science of the interactions between molecules and as the prefix bio indicates that the specific problem of biochemistry is life, now recognized as a form of super-chemistry that differs from the subject of the usual studies of chemistry* and that deals with the notion of the interrelations and self-assembly of molecules, with the concepts of their plastic genesis and of the large amount of information they carry.

The evolution from proto-biochemistry to biochemistry starts in the period in which the first conquests of chemistry were called upon to explain metabolism, after the chemical revolution, which was prepared by iatrochemistry. Ancient medicine belongs to the archaeology of medicine, and consists of empirical data and discursive developments of philosophical origin, occasionally leading to empirical active therapy, but of speculative nature and affording nothing that could be of use in the understanding of disease or in rational therapy. Nevertheless, during the 16th century, iatrochemistry was "the science" and a key to all nature. While remaining archaeologic with respect to medicine, it reached the prescientific stage with respect to chemistry. Whereas we defined biochemistry as being composed of all molecular approaches to biology, iatrochemistry, the first chemical approach to life, was in fact a biological approach to matter, with the purpose of reaching smaller values of the ratio of weight to biological activity, which became the methodological test of the iatrochemists and paved the way for the introduction of the concept of pure substances. Once chemistry had developed into a science and severed its ties with vitalist attitudes, the science of organic chemistry was in a position to develop its methods of study of the derivatives of carbon, either occurring naturally in organisms or factitious.

Organic chemistry was able to provide, by a feedback process, the background to a development of biochemistry when the inadequate concepts of biocolloidology were discarded after they had sterilized the field for half a century. If some others had a heavy responsibility in introducing the red herring of biocolloidology, several physical chemists provided the proper background to biochemical development by helping to recognize proteins as truly defined macromolecules. Studies on the structure, orderliness and functions of macromolecular compounds of several kinds, mostly proteins and nucleic acids, led to spectacular developments, owing to the introduction of a number of new methods besides the traditional chemical ones.

* It may be recalled that living organisms were outstanding organic chemists before there were organic chemists, or even men, on earth.

References p. 17

The enzymatic theory of metabolism developed in a parallel direction when the enzymes enhancing the rate of reactions were recognized as definite macromolecular units of proteins. Enzymology became a very extensive study of individual enzymes, of their chemical composition and of their mechanism of action, affording many challenging problems to the organic and physical chemist, and the reconstitution, in solution, of the steps of the many metabolic pathways of catabolism and biosynthesis. Enzymology is situated at the extreme region of the application of destructive methods, and its methodological approach has taken advantage of mincing, homogenizing and extracting tissues, and of using purified, if possible crystallized, enzymes and metabolites. The destructive character of many modern biochemical approaches has inspired apprehension in some minds. It was A. N. Whitehead, the philosopher, who posed the question to F. G. Hopkins, whether the modern biochemist, in analyzing an organism into parts, did not depart from reality to such an extent as to reach a point where his studies no longer had a biological meaning. Hopkins's answer came in the form of a lecture in which he states:

"So long as his analysis involves the isolation of events, and not merely of substances, he is not in danger of such departure". (ref. 11)

The methodology of modern biochemistry involves a critical discussion of the experimental biochemical data at hand and of their insertion into biochemical schemes and theories. Once devised, these theories have to withstand the experimental test of cells, as accomplished with the use of isotopic methods or of exacting mutants of microorganisms. This integration of data into a metabolic theory implies that there should be no contradiction between the theory and the experimental data at the level of the living material. But however productive it may have been, this purgatory of enzymology, leaving the biochemist on the threshold of the cell, could only lead to frustration until, owing to the introduction of differential centrifugation, he was finally able to enter the promised land, which since the formulation of the cell theory had remained the exclusive domain of the microscopists who at the dawn of this century, after enriching our knowledge in many domains, had exhausted the possibilities of new discovery with the help of the light microscope. It is part of the avocation of the biochemist to study the chemical structure of cells and the distribution of the effectors of biochemical activities as well as the regulation of these activities.

It had originally been the purpose of the author to limit the domain of this History to biochemistry as a fully experimental science and to start with the discovery by Buchner in 1897, of fermentation by yeast juice. But it would not have been possible to characterize the scientific development of biochemistry without contrasting this development with the archaeological era consisting of a discourse on the appearances of changing "matter" and "forces" of organisms based on observation, speculation and deductive hypothesis, and characteristic of the primitive human mind. Before being known, the modifications of matter-of-life during life were imagined.

Proto-biochemistry, to use the phrase of Needham[12], is based on such concepts as that of a preexistent, unknown microsystem, manufacturing more of itself by using matter made ready by the so-called "assimilative" changes. Proto-biochemistry deals with those very ancient and everlasting queries in the human mind such as: What becomes of food after its introduction into the body? Why is breathing associated with life and why does it cease at death? How does an organism maintain itself in the face of the tendency of its substance to destroy itself? etc.

The concept we call metabolism, though the word was not yet coined*, was one of the dominant preoccupations of the primitive mentality while it forged proto-biochemical theories to account for the steps occurring from food to circulating material, for assimilation, for decomposition and for repair, or for what we call developmental biochemistry. Such proto-biochemical discursive systems have been devised in Greek, Arabic, Indian and Chinese culture areas. The association of breathing with life, for example, has been common to all the civilizations of the Old World. As Needham emphasizes,

"The *pneuma* of the Greek was equivalent to the *prana* of India, the *ch'i* of the Chinese and later on the *ruh* of the Arabs. . . . The Indian classically counted and tried to define five sorts of *prana* and the Chinese a good many more. For a millennium and a half European thought on these matters was dominated by the Galenic theory of spirits; *pneuma*, spirit as such, coming from the circumambient air through the lungs, and combining with *pneuma physikon* or natural spirits prepared by the action of the liver on the food, to form *pneuma zotikon* or vital spirits. The former were distributed in ebb and flow from the right ventricle of the heart through the system of the veins, the latter from the left ventricle through the arteries. Similarly, under the

* The words *métabolisme* in French and *metabolism* in English, corresponding to the German *Stoffwechsel*, and used to designate the chemical reactions and the energy changes associated with them that occur in organisms, derive from the expression *metabolische Kraft* as first used by Schwann[13] in 1839. The terms *anabolism* and *catabolism*, corresponding to the ancient composition and decomposition, were coined by Gaskell[14] in 1886. The history of the use of these words has been retraced by Bing[15].

References p. 17

action of the brain, the vital spirits were transformed into *pneuma psychikon* or animal spirits, distributed to all parts of the body through the nerves. Here there was of course some connexion with the Aristotelian doctrine of 'souls', for both the natural and vital spirits were at the level of the vegetal soul (*psychē threptikē*), and the animal spirits were in the domain of the sensitive soul (*psychē aisthētikē*), while for the rational soul (*psychē dianoētikē*), because essentially mental, there was no equivalent in the world of material spirits. Ancient Chinese ideas were much more like those of the Greek than has usually been supposed, for at about the same time as Aristotle, Hsün Ch'ing and others conceived of a 'ladder of souls', *e.g. ch'i*, *shêng*, *chih* and *i*, representing a very similar ascent. From our present viewpoint, the 'souls' were but names for particular functions of the levels of living, yet we still conserve such names as 'vegetal' for the yolky poles of eggs, and 'vegetative' for parts of the nervous system." (ref. 12)

As it is to European proto-biochemistry that modern biochemistry can be historically related, we shall limit our inquiry to this domain while regretting not having the space, as well as the competence, to perform a comparative study of proto-biochemistries in the different culture areas of the Old World.

Part I of this History retraces the succession of discourses on matter-of-life and forces-of-life that were *imagined* before the chemical revolution marked the beginning of the *investigation* of the molecular aspects of life and of the evolution of a documented reduction to molecular aspects, the landmarks of which are retraced in Part II, from the isolation of proximate principles, which followed the chemical revolution, to the penetration of biochemistry into the cells. After the introduction of the methods of elementary analysis by Lavoisier, his successors, among them Fourcroy, Hallé, Prout and Liebig, fascinated by the new concept of chemical elements which had superseded the ancient concept of the four elements, were the leaders of the first episode of prescientific biochemistry, which may be designated as "analysm" and during which biochemistry was considered as a part of unified chemistry.

When this methodology was recognized as inadequate, another episode, "physiological chemistry" obtained between *ca.* 1840 and *ca.* 1880, and was characterized by another methodology, the recourse to the physiological method of vivisection, as a part of unified physiology.

Around 1880 biochemistry began to broaden its goals by calling upon a variety of methodologies, borrowed from organic chemistry, at the time having become the science of organic compounds, from physics, from genetics, from crystallography, etc.

The concern with the results of the application of a specific methodology was replaced by a concern with the specific objective of unravelling the nature, interactions and functions of the specific types of molecules found in living organisms, while organic chemistry went ahead with the study of all possible carbon derivatives, whether natural or factitious. The molecules entering into the composition of the organisms were found to conform to the physical and chemical laws of inanimate matter but, though not living, they were recognized as replete with information.

From the state of physiological chemistry as it reigned around 1880 to our own concern with cellular aspects of energetics, self-organization and self-replication, many paths of research have been followed. One of the main landmarks on these paths, the formulation of the enzymic theory of metabolism, led to the unravelling of the complicated pathways of catabolism and anabolism, the history of which will be treated in Parts III (Sources of free energy) and IV (Biosynthesis) respectively. From this study emerged the knowledge of the ways through which the cells exchange energy and matter with their environment by means of consecutive enzyme-catalyzed organic reactions, and through which the cells produce the energy-rich bonds which are the currency of the different forms of work accomplished in cells: biosynthesis, motility, active transport, electrical work, etc. This study of catabolism and anabolism revealed that the molecular level is, in the cell, the level at which the energy metabolism of organisms displays its intricate interrelations.

A trend towards the reduction of physiological and biological concepts into formulations based on the knowledge of the molecular constituents of organisms, and of their shape, structure and properties, was in the minds of the biochemists since the origin of the modern development of their science at the end of the 19th century. This trail of thought is retraced in Part V of this History, showing how it led to the concept of the cell as a self-organizing and self-replicating collection of organic molecules and of the order of life as an extremely complex molecular order.

It was for a long time believed that the coordination of the activities of organs and tissues in an organism was imposed from the whole to the parts, the cells. But when, through the brilliant achievements of molecular biology, the transmission of characters in descent became conceived as a consequence of the organization of a polymer, the approach of life was radically reversed. It was realized that the nature of an organism is not a result of statistical mechanics, leading to mean results on large numbers of molecules, but of the reproduction of an existing molecular order, conceived as extracted from disorder by differential survival. This led to the concept, as stated by Wiener, of the organism as a language and to the association of bioenergetics

References p. 17

with what may be called a *biosemiotic*, one of the parts of the general science of signs, of communication and of information, called semiology by F. de Saussure and semiotic by others, and of which linguistics is another part. The origins of the concept of organization imposed from the whole and those of the "molecular revolution" which led to the reversed approach from molecular order, will be dealt with in Part V of this History.

Biochemistry is sometimes considered as a chapter of chemistry or as applied chemistry. Of course chemical methods have been used extensively by biochemists, in the analytical phase of their work, particularly in the study of the central metabolic aspects of all cells. Moreover, many chemists have contributed to the development of metabolic studies by the synthesis of compounds proper to the accomplishment of an experimental enquiry, by the elucidation of the structure of intermediary metabolic steps, and by devising tentative chemical models for the course of metabolic pathways. The author has endeavoured to situate in their proper perspective these direct contributions of chemists to biochemical enquiry. But since it has severed its ties of servitude to unified physiology as well as to unified chemistry, biochemistry has extended its concepts and techniques in all directions appropriate to serve its specific purpose, the study of the molecular structure of the activities of cells.

Physical methods have greatly helped the progress of biochemistry through its recourse to polarography, spectroscopy, spectrophotometry, ultracentrifugation, differential centrifugation, electrophoresis, electron microscopy, isotopes, X-ray crystallography, etc. Genetic methods, such as the use of exacting mutants or the study of "inborn errors of metabolism", have also been of importance. Attitudes of the mind have been determinant in fostering biochemical progress, and it will suffice to recall the ruptures introduced by the decision of H. Buchner and M. Hahn to build a press for preparing juices of cells, of Claude to prepare a suspension of disrupted liver cells and to submit it to differential centrifugation, or of Max Delbrück to attack the study of character transmission by the bias of phage studies.

In fact, chemists cannot provide any proper methods to tackle the specific problem of biochemistry, the molecular structure of cells, and never showed any interest in this problem.

Many historians of science delight in forlorn errors and are fond of exploring the itinerary of blind alleys in the past activities of the human mind. It is not the purpose of the author to deny that such studies may have a real interest to those who study cultures of the past from an archaeological viewpoint and are interested in reviving human history. The history of proto-biochemistry will perhaps be of interest to them but it may be feared that they will be interested in this section only and will reproach the author for dealing with modern and contemporary developments. Modern history of biochemistry will on the contrary be of interest to active biochemists, as this domain is still far from being dead, and remains the background of modern research. It is of interest to biochemists, as well as to historians, to become acquainted with the nature of ruptures and transitions at the level of which periods of sterility are suddenly replaced by trails of research leading to great rewards. While the philosopher may be interested in such aspects of genetic epistemology, as ways of reaching what he calls "truth", the biochemist may be interested, more humbly, in discovering the sources of the trails he is following. The selection of the starting point of the fruitful enquiry, the judicious choice among all possible techniques and concepts, the replacement of wasting efforts by an acceleration of discovery and the torrent-like accumulation of answers to queries, and of the flux of new paradigms: all these aspects are alive in a young science, and, far from serving idle curiosity, they have a stimulating virtue. Why did we put the History at the end of *Comprehensive Biochemistry*? The purpose is to link the History with the histories of the recent developments the different authors of the Treatise have mentioned in their presentation and, with their expert collaboration, to provide the reader with a history reaching up to the present time.

Modern biochemistry is mechanistic and is a part of soulless science. But in proto-biochemistry and in prescientific biochemistry, the philosophical and religious life-matter problem, absent from the biochemical conceptual system of our days, has repeatedly appeared and has influenced primitive concepts of "matter" and "forces" as they were phenomenologically grasped at the level of living organisms.

In a very documented and thoughtful book, Hall[16] has dealt with the subject of the relations and interactions of the scientific concepts of life and matter, a subject of a philosophical and religious nature. When its incidence

References p. 17

was present in our developments we have referred the reader to Hall's book for a more detailed treatment. Hall recognizes in the history of the formulations of the life–matter relations, several distinct varieties. Life has been thought of as an actual variety of matter, as a material differentiation of the Cosmos. According to Aristotle, Anaximenes identified life-as-soul with air, and Heraclitus with fire (*life as identical*). It is sometimes difficult to differentiate this formulation from the sort of life-matter relation of immanence, *i.e.* matter related to life in the way that extension or mass is (*life as immanent*). The formulation of life as immanent was revived in the 18th century by Buffon, for instance, and in the 19th century by Haeckel. Life-as-soul, for Plato, is a distinct non-material entity bonded to certain objects and "inducing in them a characteristic ensemble of behaviour (*life as action*)"[16]. The relation here is one of imposition (*life as imposed*). Occasionally, life has been considered as "a special arrangement of matter that permits it to behave in a lively fashion". Aristotle introduced this formulation when he equated life-as-soul with form (*life as organization*).

Life has also been considered as emergent, *i.e.* as "an ensemble of distinctive activities that certain objects (organisms) engage in solely because of their material constitution"[16] (*life as an emergent consequence of organization*). Epicurus used this formulation, insisting that it was absent from single atoms. Since the Renaissance this formulation has progressively displaced all the others.

"*Life as identical*" and "*life as immanent*" have in the past been designated as *hylozoic*, while "*life as imposed*" has been known as *vitalist*, and "*life as emergent*" as *mechanistic*.

The development of biochemistry has not followed a straightforward path to accepted paradigms, and many blind alleys were at times abandoned after long sterile exploration and passed out of ken. The author has tried to take them into account at the stage of their formulation, when their consideration affords useful insight into the nature of the origin of these erroneous trends. On the other hand, in order not to make the number of references overwhelmingly large, no mention has been made of a number of publications of minor interest, or secondary to the formulation of ill-fated theories, to which an enormous amount of literature has sometimes been devoted. The history of the elucidation of the nature and composition of natural chemical compounds belonging to the theoretical and methodological domain of the organic chemist, but essential for the progress of biochemistry, has, in spite of its great importance and interest, been omitted with regret.

The author owes a great debt of gratitude to his predecessors in the field of the history of biochemistry. The regretted Fritz Lieben, author of a *Geschichte der physiologischen Chemie*, published in 1935, divides his book into two parts. The first, extending from antiquity to the 19th century, is written in the biographical mode. The second appears as a prehistory of the development of the molecular basis of the *functions* of the physiologist, a subject which was still in its infancy in 1935. Lieben writes in the dark ages of biocolloidology, at a time when Neuberg's theory of glycolysis was still standard, when the enzymes and other proteins had not yet been recognized as definite molecular entities, when the concept of the energy-rich bond had not yet been defined, and when nothing was known on the pathways of biosynthesis. As he was more interested in natural substances themselves than in their fate in the organism and in its chemical structure, Lieben devoted many pages of his treatise to problems relating to the organic chemistry of natural substances. He nevertheless inserted chapters on muscle, blood coagulation, blood, bile, urine, etc. in harmony with the analytical preoccupations of the "physiological chemists" of the time.

The author has also repeatedly quoted from such books as Partington's *A History of Chemistry*[18], Mendelsohn's *Heat and Life*[19], and Hall's *Ideas on Life and Matter*[16], but it will not excuse the reader from perusing them as invaluable sources of additional information and inspiration.

The late Otto Warburg used to recall Maxwell's plea to Ampère:

"If you have built up a perfect demonstration, do not remove all traces of the scaffolding by which you have raised it." (ref. 20)

It would certainly have been of invaluable interest to those who are seeking to determine the origins and development of the concepts of biochemistry if more among the pioneers of the art had taken the pains to recall the nature of the scaffolding of their discoveries as it has been done in Warburg's *Schwermetalle als Wirkungsgruppen*[21], Keilin's *History of Cell Respiration*[22], Du Vigneaud's *A Trail of Research*[23], Kalckar's *Biological Phosphorylations*[20], Watson's *The Double Helix*[24], Lipmann's *Wanderings of a Biochemist*[25], Dorothy M. Needham's *Machina Carnis*[26], the essays collected under the title *Phage and the Origins of Molecular Biology*[27] and the Prefatory Chapters published since 1956 in *Annual Reviews of Biochemistry*.

References p. 17

While based on the original literature, such a historical treatment as this was greatly helped by the perusal of the historical introduction to the original publications, as well as to books and historical review papers written by specialists in the different fields of study. To the authors of such reviews to which due credit will be given in the text, the author is greatly indebted.

REFERENCES

1 F. Hoppe-Seyler, *Z. Physiol. Chem.*, 1 (1877).
2 H. A. Krebs, W. N. Aldridge, K. S. Dogson, S. R. Elsden, G. A. D. Haslewood, A. P. Mathias, D. C. Phillips and R. M. S. Smellie, *Biochemistry, "Molecular Biology" and Biological Sciences*, The Biochemical Society, London, 1969.
3 T. S. Kuhn, *The Structure of Scientific Revolution*, Chicago, 1962.
4 A. L. Lehninger, *Biochemistry. The Molecular Basis of Cell Structure and Function*, New York, 1970.
5 F. Jacob, *La Logique du Vivant*, Paris, 1970.
6 G. Canguilhem, *Etudes d'Histoire et de Philosophie des Sciences*, Paris, 1968.
7 H. F. Blum, *Am. Scientist*, 49 (1961) 474.
8 L. Belloni, *Progr. Biochem. Pharmacol.*, 1 (1965) 5.
9 G. Bilancioni, *Il Valsalva (Riv. Mens. Oto-Rino-Laringol.)*, 4 (1928) 85.
10 J. T. Edsall, in G. R. Barker (Ed.), *The Training of Biochemists*, The Biochemical Society, London, 1966.
11 F. G. Hopkins, *Science*, 84 (1936) 258.
12 J. Needham, in J. Needham (Ed.), *The Chemistry of Life*, Cambridge, 1970.
13 T. Schwann, *Mikroskopische Untersuchungen über die Übereinstimmung in der Struktur und dem Wachstum der Thiere und Pflanzen*, Berlin, 1839.
14 W. H. Gaskell, *J. Physiol. (London)*, 7 (1971) 1.
15 F. C. Bing, *J. Hist. Med.*, 26 (1971) 158.
16 T. S. Hall, *Ideas of Life and Matter*, 2 Vols., Chicago, 1969.
17 F. Lieben, *Geschichte der physiologischen Chemie*, Leipzig and Vienna, 1935. [reprinted by Olms, Hildesheim, 1970].
18 J. R. Partington, *A History of Chemistry*, Vols. 1/I, 2–4, London, 1961–1970.
19 E. Mendelsohn, *Heat and Life*, Cambridge, Mass., 1964.
20 H. M. Kalckar, *Biological Phosphorylations*, Englewood Cliffs, 1969.
21 O. Warburg, *Schwermetalle als Wirkungsgruppen von Fermenten*, Berlin, 1946.
22 D. Keilin, *The History of Cell Respiration and Cytochrome*, Cambridge, 1966.
23 V. du Vigneaud, *A Trail of Research*, Ithaca, 1952.
24 J. D. Watson, *The Double Helix*, New York, 1968.
25 F. Lipmann, *Wanderings of a Biochemist*, New York, 1971.
26 D. M. Needham, *Machina Carnis, The Biochemistry of Muscular Contraction in its Historical Development*, Cambridge, 1971.
27 J. Cairns, G. S. Stent and J. D. Watson (Eds.), *Phage and the Origin of Molecular Biology*, Long Island, 1966.

Foreword to Parts I and II

The author wishes to express his gratitude to several colleagues and friends for their valuable assistance in discussing or in reading parts of the manuscript, and for their valued suggestions, and particularly to A. Claude, A. G. Debus, C. de Duve, M. D. Grmek, F. L. Holmes, R. Joly and G. Palade.

For permission to reproduce copyright material the author expresses his indebtedness to the following: W. H. Brock, A. Claude, Ch. McC. Brooks, C. de Duve, G. Heym, F. L. Holmes, H. S. Mason, E. Mendelsohn, J. Needham, R. Olby; The Cambridge University Press, London; David Higham Associates Ltd., London; Harvard University Press, Cambridge, Mass.; Heinemann Educational Books Ltd., London; The Society for the Study of Alchemy and early Chemistry.

For the illustrations, the author also expresses his gratefulness to those who helped him and whose kind contribution is acknowledged in the list of illustrations (p. XV).

Liège, June 1972 M. Florkin

Part I

Proto-Biochemistry

Chapter 1

Ancient Greek Proto-Biochemistry

1. Proto-biochemistry of the presocratic thinkers

For the Greek presocratic philosophers, life appeared as action (nutrition, locomotion, generation) and as soul (ability to set things in motion, assimilation of food, perception, emotion, reason, etc.), and when they started analyzing life-as-action, they did not have to give up life-as-soul.

The Greek presocratic philosophers[1] restated and elaborated many ancient concepts of greater antiquity which they had obtained from what was at the time the Persian Empire. Among these ideas was the concept of microcosm and macrocosm. They admitted that the knowledge of one of these would afford knowledge of the other. Instead of starting from the macrocosm and indulging in astrology, like the Babylonians, they considered man as the centre of the system. A general proto-biochemical aspect of ancient Greek natural philosophy derives from the common observation of the maintenance of the body size and weight in spite of regular ingestion of food and regular excretion of faeces and urine. This maintenance of a proper balance between receipt and expenditure in what was considered to derive from a process of waste and repair of the body was extended to the whole of the Cosmos.

The process was considered as being an addition or removal of minute particles. The processes of "*composition and decomposition*", "*waste and repair*", also later called "*dynamic permanence*" by Regnell[2] had, according to Heidel[3], been formulated by Alcmaeon at the beginning of the 6th century B.C.

For the presocratic philosophers, there was no other cause of things than matter itself. According to ancient tradition, Thales (probably 624–548 B.C.) considered water to be the primeval substance of everything, the "*principle*", that from which everything comes, including living organisms.

References p. 50

The origin of the concept is related to common sensorial knowledge: man arises from sperm and subsists on moist nourishment. For Thales, the whole universe was alive ("everything is full of gods"). His Milesian compatriot Anaximander (611–545?) identified the primary matter with "*apeiron*": one and undefined, eternal, un-ageing, indestructible, in eternal motion. From *apeiron*, many worlds were continually arising, to return to it again.

Anaximander's pupil, Anaximenes (586–528?) identifies matter with air. While the *apeiron* of Anaximander was *that from which* everything, including organisms, came to be, as well as *that into which* everything resolves, the air has not only these qualities; for Anaximenes, air is also *that of which* everything consists. Every substance is nothing but a modification of air by condensation or rarefaction. Rarefied air is fire, while condensed air successively becomes water, earth and finally stone. Boundless, inexhaustible and in continuous motion, it is the soul (psyche) of the universe and the principle and the condition of life, as water had been for Thales[4].

Heraclitus identified primary matter with fire, with the connotations later given to *vital heat* (*innate heat*). Condensed, it successively became moist, water, earth. Rarefied earth became water. The idea according to which the soul is a kind of fire, later accepted by Democritus, is already present, according to Aëtius, in the teachings of Heraclitus, Hippasus and Parmenides. According to tradition, Hippasus was a member of the Pythagorean school. In its earlier cosmology, this school believed that the world was derived from a fiery material (seed). In the philosophy of Heraclitus, a *struggle between opposites* plays an important role. The idea was an ancient one, already found in Anaximander's philosophy (from the first cause, *apeiron*, arose heat and cold, dry and moist)[4], and it will remain an ever-present theme in proto-biochemistry as well as in biochemistry.

Heraclitus developed a dynamic concept according to which all elements are continuously following the downward and upward paths of Anaximenes. For Anaximenes, air and fire were interchangeable, air following a downward path to water and earth by condensation, and an upward path to fire. The conception of Heraclitus, in his endeavour to explain the macrocosm by man, the microcosm, led him to the first conception of metabolism, the system being composed of fire, water and earth, ever changing into one another[5]. Nevertheless, the body as a whole does not appear to change.

For Heraclitus, such phenomena as sleep or wakefulness resulted from minor oscillations in the interconversion of the basic elements. Consciousness had its source in fire, sleep resulting from a decrease of its amount.

A certain equilibrium between fire and water resulted in the existence of the soul (life) and death resulted if either became predominant[5].

The philosophers mentioned above, who are sometimes called "physiologists" owing to their primary interest in living organisms, have also been called "monists" as they assumed the existence of one kind of matter and explained how all things derived from it. On the contrary, Empedocles (490–435? B.C.), who based his theory on cosmology (macrocosm), postulated the existence of different kinds of matter, irreducible to one another. He has therefore been called a "pluralist". This development may have followed the rejection by the philosophers of Elea (Parmenides, Zeno) of the existence of a void, this rejection leaving no possibility for condensation or rarefaction.

The teaching of Empedocles may be understood from the following fragments of his treatise *On Nature*:

"Hear first the four roots of all things: bright Zeus, lifegiving Hera and Aidoneus and Nestis who moisten the springs of men with their tears. And a second thing I will tell thee: There is no origination of anything that is mortal, nor yet any end in baneful death; but only mixture and separation of what is mixed, but men call this "origination." (translation cited by Collingwood[6])

Empedocles's pluralism is based on four autonomous kinds of matter: earth (Aidoneus), water (Nestis), air (Hera) and fire (Zeus), the relation of which with common sensorial knowledge, and at the same time with religious personifications of natural events, are obvious. The "roots" or "elements" of Empedocles produce the variety of material objects. The "elements" are themselves universal, eternal and unchangeable while changes of the material objects result from their mixing in different proportions. The elements of Empedocles are passive, and in order to realize or to separate the different mixtures, he introduces the concept of two active principles, described as "forms of matter" and to which he refers as "love" and "strife". When fire predominates the animal flies upwards like the birds. If fire and earth are separated by "strife", this results in death.

"But these (elements) are the same, and penetrating through each other they become one thing in one place and another in another, while ever they remain alike. In strife all things are endued with form and separate from each other, but they come together in love and are desired by each other. For from these (elements) come all things that are or have been or shall be; from these there grew up trees and men and women, wild beasts and birds and water nourished fishes; and the very gods, long-lived, highest in honour." (translation cited by Collingwood[2])

All existing objects result from different mixtures of the four primary kinds

References p. 50

of matter, and Empedocles describes the concept in the following model:

"Just as when painters are elaborating temple-offerings,—men whom wisdom hath well taught their art—they, when they have taken pigments of many colours with their hands, mix them in due proportion, more of some and less of others, and from them produce shapes like unto all things... So let not the error prevail over thy mind, that there is any other source of all perishable creatures that appear in countless numbers." (translation cited by Lejewski[4])

For Empedocles, the organism of man and all other terrestrial bodies are the result of the mixture (not fusion) of the same four elements, true elements unchangeable and ultimate "chemically". As Collingwood underlines,

"Empedocles seems to hold that the elements merge their separate natures so as to become one thing in one place and a different thing in another. This is quite different from holding that one element has four different forms, for it implies that many new things are composed from the elements and yet the components remain somehow what they were but not in their original form. It not only increases the number of basic stuffs, but it also proposes that by their penetrating one another in various proportions, many other elements of things come to be. In this way, Empedocles took a long step toward squaring the intellectual account with the sensory evidence that attests to a multitude of truly diverse things in the universe." (ref. 6)

According to Empedocles, flesh and blood are formed of equal weights of the four elements, while sinews are composed of fire and earth, and bones from one half fire and one fourth each of earth and water.

Nutrition was accounted for in Empedocles's philosophy which created the long-lasting myth of *direct assimilation*. According to him like attracts like and each part of the organism attracts to itself certain suitable materials to construct it further. The assimilated materials occupied the place they did because they were of the right size to fit in the pores of the similar portion of the body. Blood circulated from the heart (the organ of consciousness) to the surface of the body (where air was taken in through the pores) and back, expelling and drawing in air alternately[5,7].

The theory of the four elements had a tremendous success and it can be classified among the theories that lasted for an exceptionally long time. The four elements are still present in the poems of Keats. Even nowadays it is common to read that the "elements" are threatened with destruction. A few years ago, the painter André Masson entitled one of his paintings "The four elements". This everlasting popularity evidently is the result of the antiscientific nature of the theory and of its direct appeal to realism and sensorial apprehension accessible to all individuals, even those incapable of abstract thinking.

Trivial experience and common knowledge were embodied by it. The quality of stability and resistance is substantiated in "earth", constituting either a protection or an obstacle. This substantiation is followed by an animist move.

The element fire manifests itself in common life as what breaks down not necessarily by flame. It is in all antiquity linked with lysis, "coction", with decay and with spontaneous generation. It is in relation with the idea of soul, in an extreme valorization of the concept which makes it the source of all functions in the organism.

The element water is obviously part of generation, as sperm is fluid and because fluidity appears everywhere in nature, in rivers, sea and lakes as well as in blood, tears, milk, saliva or in wine. Man drinks fluids and passes urine. The element water is associated with all the shades of natural waters, impetuous or stagnant, the matrix of all things.

The element air is essentially light and mobile, calm or animated by storms and winds, and striking man with occasional disasters.

Empedocles had gathered the components of his system from many sources. Zeller[8] considers his philosophy as a reconciliation between the permanence of Parmenides and the generation, decay and motion of Heraclitus. In a unitary vision of the world, of birth and death and of the mysteries of nature, he combined elements of the most diverse origins: the cosmogonies of Homer and Hesiod, Orphic mysticism as well as his experience as a physician and naturalist[9]. Of the extensive poetical work of Empedocles, few fragments have been preserved. These, together with the comments and citations of ancient writers, have recently been assembled by Bollack in a scholarly edition[10].

The material pluralism of Anaxagoras (500?–428) who was older than Empedocles, but published his treatise after Empedocles's poem had been read for some time, accepted an unlimited number of kinds of substances and their infinite divisibility. In everything, according to him, there are seeds of everything else. These "seeds" were not perceptible by the senses but they could be known by reason. Anaxagoras considered that in the air and in the soil all the components of a tree are present, and in the food all the components of the body are contained. From the food, in the form of bread and water, derive a selection of its components for such different parts as blood, flesh, or bone. The fundamental substance is associated in each case by the action of the *nous*, or mind, a substance that is not mixed with the other and that is endowed with a superior power. The substance is the basis of setting the world in motion, and it is a prime mover, a *causa efficiens* in the same way as the soul is considered by Aristotle. According

References p. 50

Plate 1. Hippocrates.

to Anaxagoras, life first arises from moisture, earth and heat. For Anaxagoras, the air contains the seeds of all things, and when the seeds were brought to the soil with the rain, the plants resulted. He believes that all the components for the nutrition of trees are contained in the air and in water[11]. As underlined by Lejewski[4], Anaxagoras is primarily interested in the concept of matter as *that from which* all things come to be and *that of which* all things consist. The concept of *that into which* all things are resolved does not seem to play a role in his philosophy.

By the end of the 5th century B.C. the Greek thinkers had developed a number of concepts which had a definite impact on the way of conceiving the changes of matter and of "forces" involved in organisms. These concepts were the idea of several elemental substances, the properties of which were based on certain physical qualities, and the idea that an equilibrium existed between these properties.

The doctrine of balance was first applied to organisms by Alcmaeon of Croton (who lived about 500 B.C.). The older doctrine of pairs of opposites he combined with the harmony of the Pythagoreans. The opposites of which he supposes the existence, are not substances but "powers", and he conceives many of them (hot, cold, moist, dry, bitter, sweet, etc.). Disease results from the predominance of one of the powers. Alcmaeon's equilibrium was adapted by Empedocles to the material substances he adopted as "roots".

2. Hippocratic concepts of metabolism

The so-called *Hippocratic collection*[12], comprising about 60 medical writings, dates from the last part of the 5th century or from the first decades of the 4th. Some are a little more ancient and a few others were composed much later[13]. The medical aspect of these treatises is outside our interest. It has been the subject of extreme overvalorizations in certain medical circles, an aspect which has been analysed by Joly[14]. Of Hippocrates himself, who was one of the leading figures of the school, we know very little. He was probably born in 460 and he was from Cos. It is impossible to determine his part as an author in the writings of the treatises forming the *Hippocratic collection*. The doctrine of the balance of the four humours is in fact the first metabolic theory based on observation. Body fluids have been known by the healing profession from time immemorial. No scientific or philosophic knowledge is required to see a bleeding wound, or a running

References p. 50

nose. The body fluids are taken into account in the ancient medicine of the Babylonians or of the Egyptians. They considered blood as the main body fluid but they also knew bile and mucus.

A great amount of confusion arises from an identification of the four "hippocratic" humours (blood, yellow bile, black bile, phlegm) with body fluids. The body fluid blood is not the hippocratic humour blood. On the other hand the four hippocratic humours are not directly related to the four elements of Empedocles, which are taken as granted in the natural philosophy of the time. The humours are related to the concept of the four qualities, foreign to presocratic philosophies. The four qualities and the two oppositions warm-cold and dry-moist are the fundamental qualities in the hippocratic theory, which does not mean that these oppositions did not occasionally appear among many couples of contraries as components of older doctrines in which such couples of contraries frequently appear (love and hate in Empedocles, light and night in Parmenides, etc.) in a struggle between opposites. The four hippocratic humours, or *chymoi* (blood, phlegm, yellow bile, black bile) are related to the two fundamental couples of contraries of qualities warm-cold and dry-moist, already mentioned by Anaximander. Phlegm is the coldest constituent and predominates in winter, while blood, which is moist and warm, is associated with spring. Yellow bile, dry and warm, is associated with summer, while black bile, dry and cold, is associated with autumn. The theory of the balance of the four humours is the first analytical metabolic theory. In this system, for instance, the body fluid blood is revealed by the observation of the phenomenon of coagulation to be composed of four constituents (humours). The bottom, black part of the clot (methemoglobin) was the humour black bile, while the red part (red blood cells) was the humour blood, and the serum the humour yellow bile. The fibrin was the humour phlegm. The theory of the balance of the four humours, themselves unalterable, considers that only their relative proportions may vary in a body fluid or in a solid part of the body. Among extant writings, the most ancient in which the theory of the four humours is formulated is the hippocratic treatise *The Nature of Man*, generally attributed to Hippocrates's son-in-law, Polybes. In *The Nature of Man*, which has been the subject of a critical analysis by Joly[14], the first eight chapters concern the Empedoclean doctrine of the four elements and expose the arguments against the monist theories. These objections also apply to physicians who pretend that man is only blood, or only bile, and a few only phlegm. But, writes the author in Chapter II,

... "I hold that if man were a unity he would never feel pain, as there would be nothing from which a unity could suffer pain. And even if he were to suffer, the cure too would have to be one. But as a matter of fact cures are many. For in the body are many constituents, which, by heating, by cooling, by drying or by wetting one another contrary to nature, engender diseases; so that both the forms of diseases are many and the healing of them is manifold." (translated by Jones[12])

Later, in Chapter IV:

"The body of man has in itself blood, phlegm, yellow bile and black bile." (translated by Jones[12])

And, in Chapter VI, opposing the view according to which as a man who has his throat cut bleeds to death, it must be concluded that blood composes the soul of a man, the author writes:

"Yet first, nobody yet in excessive purgings has vomited bile alone when he died. But when a man has drunk a drug which withdraws bile, he first vomits bile, then phlegm also. Afterwards under stress men vomit after these black bile, and finally they vomit pure blood. The same experiences happen to those who drink drugs which withdraw phlegm. First they vomit phlegm, then yellow bile, then black and finally pure blood, whereas they die. For when the drug enters the body, it first withdraws that constituent of the body which is almost akin to itself, and then it draws and purges the other constituents." (translated by Jones[12])

In the medical developments of the theory, the characteristics of the traditional approach of the clinicians are already recognizable. As based mainly on observation, medical science remains even today a retarded science. Chemistry suffered the same drawback for a long time, as the chemical object was difficult to grasp in primitive science. Biochemistry also remained proto-biochemistry until the organic chemists were able to identify the substances entering into the constitution of the organism. The retarded state of scientific pathology has long depended on the same obscurity as the pathological object remained occulted by the taste for the immediate sensible reality, common among the prescientific clinicians, prone to an overdetermined method of thought, was the mixing of the most heteroclitous observations, in which everything may be the cause of everything, and masking the real determination. To break with the immediate reality of the patient was possible only when the nature of the organism was better known. In any case the handling of the diseases begins with observation, an aspect on which it is true that the Hippocratics devoted much attention, even if their medical theories have been, and still are, greatly overvalorized.

The hippocratic approach, however primitive it may be in that respect, also has the great merit of being an analytical pre-biochemical approach.

References p. 50

Among the components of the body fluid blood, the two that are recognized as yellow bile and phlegm are still respectively considered by us as individual constituents under the names of serum and fibrin. However primitive their proto-biochemical analysis may have been, the Hippocratics must be recognized as the first who practised this analysis, which may be considered as a first attempt at the isolation of the "proximate principles", as they were to be called in the 19th century.

The influence of Anaxagoras is recognizable in the hippocratic treatise *Regimen*. According to Joly[13,15] this treatise dates from around 400 B.C. The balance of fire (hot and dry) and water (cold and moist) causes continual changes in the body:

"Such is the nutriment of man, one part pulls, the other pushes; what is forced inside comes outside." (translated by Jones[12])

The structure of the organism comes from food:

"Thus food and breath by movement and the action of fire are dried and solidified, hardening around the outside. This traps the fire inside, and no more nourishment can be drawn, while breath cannot be expelled through the hard outer layer. The available interior moisture is consumed; the interior part becomes compact and solid, and, in fact, becomes bones and sinews. In softer tissues the fire which is enclosed can break out, leaving passages for the breath and to supply nourishment. These are the hollow veins. Between these, and what is left over from the water becomes compact and is converted to flesh." (translated by Jones[12])

In the treatise on *Nutriment*, it is stated that food

"increases, strengthens, clothes with flesh, makes like, makes unlike, what is in the several parts, according to the nature of each part and its original power." (translation by Jones[12])

And in the *Regimen*:

"Now the things that enter must contain all the parts. From what of which no part were present would not grow at all, whether the nutriment that were added were much or little." (translation by Jones[12])

Concerning the preparation of the food for assimilation by the process of digestion, the *Nutriment* states:

"Power of nutriment reaches to bone and to all parts of bone, to sinew, to vein, to artery, to muscle, to membrane, to flesh, fat, blood, phlegm, marrow brain, spinal marrow, the intestine and all their parts; it reaches also to heat, breath and moisture." (translation by Jones[12])

As pointed out by Winslow and Bellinger, the last words of this citation:

"carry us over from the concept of food as a building-material to food as a source of energy." (ref. 16)

In the treatise of the Hippocratic Corpus, the importance of the blending of opposites (qualities, substances, humours) is stated. This blending corresponds to the concept of *pepsis* (generally translated by coction or concoction). The Greek verb is *πεπτω* from which *πεψισ* is derived. The corresponding Latin words are *concoquo* and *concoctio*. It is not easy to define the meaning of coction. For Leicester

"It was the method by which the diverse elements in the food or in the body were brought into the intimate blending required for the normal functioning of the body for perfect health." (ref. 5)

For Jones (in the introduction of *Hippocrates*, Vol. 1) coction is the

"action which so combines the opposing humours that there results a perfect fusion of them all. No one is left in excess so as to cause trouble or pain to the individual." (ref. 12)

The concept corresponds to the idea of balance already present in Alcmaeon's philosophy. Coction, accomplished by heat, occurs at all times and thickens body fluids. Fluids become thinner if coction is imperfect; this is the cause of disease. On certain days of the disease, there is a restoration of normal concoction (*crisis*).

3. The atomists

Atomism is best known to us in the form of Epicurean natural philosophy, through the fragments of writings of Epicurus, among which are his *Letters to Herodotus and Pythocles*, as well as through other writings accompanying these letters in their publication by Bailey[17].

Epicurean philosophy is also widely known through Lucretius's *De Natura Rerum*[18]. The natural philosophies of Democritus and of Leucippus are less well known.

According to ancient beliefs, death implies that the soul leaves the body. The atomists believed that the *psyche*, the soul, left the body at death. There must have existed a host of theories concerning *psyche*. For those philosophers defined as "atomists", who considered the organisms as being composed of mobile *compact* particles, the atoms, separated by completely *empty* space, the soul was a substance formed of a number of atoms of a special nature, small, round and smooth, and able to move rapidly in the empty spaces, thereby setting the body in motion through their impact with other atoms. (This concept is attributed by Aristotle to Theophrastus, to Leucippus and to Democritus, and it is clearly formulated by Epicurus.) As the high mobility of fire is well known, and as it confers mobility to water by bringing it to boil, fire, considered as a substance, must be com-

References p. 50

posed of small round atoms like the soul. The soul is therefore considered by Leucippus and Democritus as being a kind of fire. The theory of Epicurus was more complex, as he considered the soul

"as a body of fine particles distributed throughout the whole structure and most resembling wind with a certain admixture of heat" (letter to Herodotus). (translation by Bailey[17])

Other texts also show that Epicurus considered the soul as consisting of four kinds of substance: "heat", "air", "wind" (or substances rather similar to these) and a "nameless substance". These four soul elements are also mentioned by Lucretius*. Respiration therefore accomplishes two functions: to offset the loss of soul atoms and to replace lost soul atoms.

Concerning the mechanism of the maintenance of the dynamic permanence, the atomist could not help but follow the teachings of trivial observations and notice that an organism has its characteristic ways of absorbing food and processing it into its own substance.

As Regnell states (essentially from Lucretius):

"For example, when an animal eats grass, the grass is transformed during its passage through the body into muscle, bone, blood, etc. The organization of atoms which thus takes place presupposes that the food is processed in the body. This is achieved for example, by chewing. However the soul and the fire atoms are here of decisive importance in the same way as bodily warmth in spontaneous generation. Due to their rapidity, smallness and spherical shape, fire atoms penetrate easily through matter. They are thus able to break food down into its smallest components. The various organs, tissues and fluids in the body then receive their necessary additions in the form of new atoms which pass into their structure." (ref. 2)

For the atomists, metabolism was an exchange of matter.

If the atomists considered that the soul leaves the body at death, they did not believe in the eternity of the soul. On the contrary, they believed it is mortal and cannot exist outside the body. The body is the container without which the soul cannot be held together. The small atoms of the soul are submitted (as is the fire in the world at large) to a strong pressure, but during respiration the atoms which have left the body are replaced by new ones contained in the air. When breathing ceases, the pressure from outside is not counteracted, and the soul atoms are squeezed out of the body. This view of Democritus is rendered by Aristotle in *De respiratione*.

According to the atomists, the small round atoms responsible for the body's driving force are the source of animal heat. Thereby a close relationship is established between heat and atoms in rapid motion. When the dead body cools, it is the result of the rapidly moving atoms (soul) leaving

* On these aspects of the soul according to the atomists see Bailey[17], Regnell[2] and Hall[19].

the body (small soul atoms gliding through the "pores" of the body). This implies that the introduction of the small rapidly moving atoms was necessary for spontaneous generation (as solar heat is necessary for the rebirth of life in spring, and birds' heat for the hatching of eggs: simple current observations, tendency to generalizations).

Nevertheless, the duration of life is dependent on the conservation in the body of a supply of soul atoms and of the maintenance of the cooperation of body and soul (both composed of atoms) in a unit, the organism. The organism continuously loses atoms from its interior and its surface, and new ones are added. A new atom may take the place of an old one, and in a certain sense, the body is not changed (dynamic permanence)[2]. The atomists were not free from the sin of animism and they undoubtedly described the world in certain aspects as a whole, as a kind of organism, but they did not consider the world as a whole as a living being and as having a soul. Epicurians did not consider plants as living in a true sense, *i.e.* as having a soul. The growth and motion of plants was not considered by them as due to a soul. (On this important aspect see Bailey[20].)

4. Plato's views on metabolism

It should first be recalled that in the cosmological system presented by him in the *Timaeus*, Plato denies reality to matter and states that, for him, the world is an image, not a substance. In constructing a picture of the universe, Plato does not start from the results of an analysis but starts from below, from the elementary indivisibles, and to underline its conjectural nature he describes what he defines as a story, as a myth.

The reader will derive great profit from perusing the lucid commentary of the *Timaeus* by Cornford[21], to which the following pages are greatly indebted.

One of the most interesting features of the *Timaeus* is that it shows how Plato was preoccupied by one of the ever-present themes of Science, the way by which the order of life takes over the disorder of inanimate nature. He resolves it as the action of a Demiurge. Plato's Demiurge, the divine artificer, is not the omnipotent creator of the Jewish–Christian religion. On the other hand the concept of natural law, as a system of causes and effects, is entirely foreign to Plato's intellectual structure.

The Demiurge, as far as he can, introduces order into chaos. While the God of Moses, by will, introduces matter into order, Plato's Demiurge

References p. 50

Plate 2. Plato.

limits his action to the domain of what is, by nature, possible and chooses the best of his contributions within the inherent possibilities depending on the nature of the materials. He takes over a chaos of disorderly motions and powers and reduces it to order, as much and as well as is possible. The Demiurge is a symbol, akin to the "personified Nature" of Aristotle, aiming at a purpose.

The order introduced into chaos by the Demiurge consists of the "intelligible Living Creature", a system of Forms, containing four families of species, the "heavenly gods" (stars, planets, earth), the birds, the fish and the animals living on land. The world appears as a single living body, in opposition with the theories of the atomists who believed in a plurality of existing worlds.

The four families of living creatures correspond to the four regions of fire, air, water and earth. The heavenly bodies are the gods to whom the Demiurge delegates the production of the three other categories of creatures, a recognition of the contribution of those heavenly bodies, and of the sun in particular, to the generation of life on earth.

Plato recognizes indivisible elementary constituents. But, in contradiction to the views of the atomists, whose indivisibles were solids, the indivisibles of Plato are planes and the figures, limited in number, derived from them. Plato's Demiurge has assigned to the four elements of Empedocles to which he gives the nature of qualities, not substances, the figures of the regular geometrical solids, thereby introducing order into the field ruled by irrational powers, the four shapes of the regular solids assigned to the four elements being the best, aesthetically as well as with respect to their reciprocal associations.

The triangle, as the surface contained in a minimal number of straight lines, is considered as the irreducible element of any rectilinear plane face. Any triangle can be divided into two right-angled triangles, both isosceles or both scalene, or one scalene and one isosceles. The four elements, Plato assumes, were cut out of space by two right-angled triangles, very small and no more visible to us than the double helix of DNA. One kind of these triangles (α in Fig. 1) is equilateral and made by dividing a square into four equal parts. The other kind (β in Fig. 1) is obtained by cutting an equilateral triangle into six equal parts, one side of each part being half as long as the hypotenuse. From combinations of β triangles we may construct the tetrahedron or fire (4 equilateral triangles), the octahedron or air (8 equilateral triangles) and the icosahedron or water (20 equilateral triangles). From

References p. 50

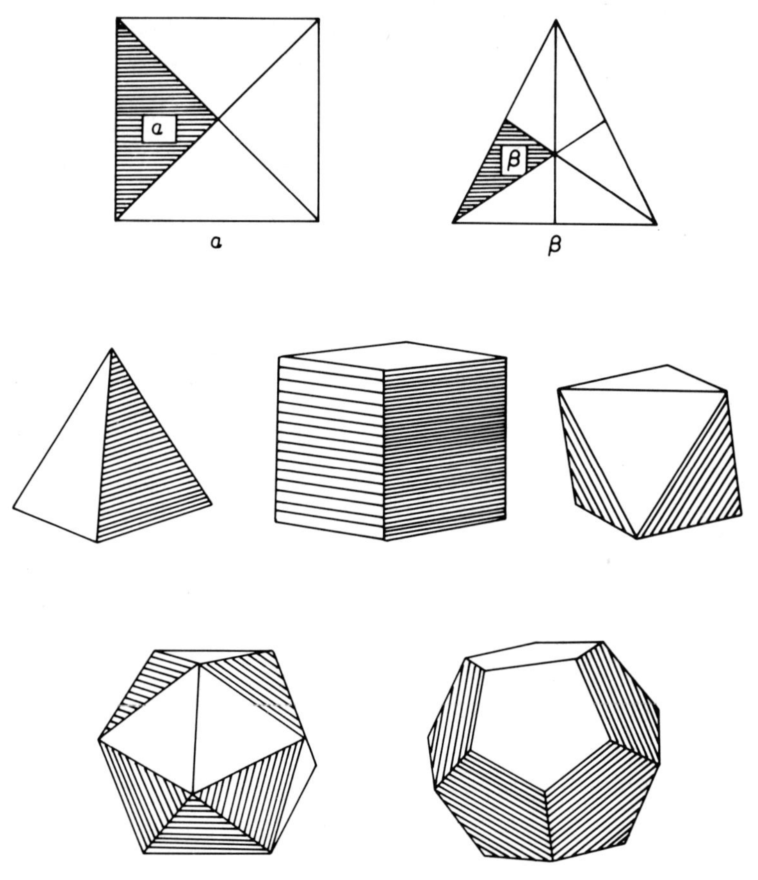

Fig. 1. The two platonic triangles and the five platonic bodies.

combinations of α triangles, the hexahedron (cube, six squares) or earth is constructed. The four "regular solids" are figures with all sides equal and all angles equal. There exists a fifth "regular solid", the dodecahedron, constructed from 12 regular pentagons, and which Plato considers as the symbol of the whole, nearly spherical universe. The "regular solids" are of Pythagorean origin, but it was Plato who identified with them the four elements of Empedocles.

It must be stressed that the figures are not considered by Plato as the actual shapes of existing particles but as the perfect mathematical types, only imperfectly copied and existing in different grades of size.

The transformation of a primary body into another results from the breakdown of the solid form of the particle (cube number) into plane faces (square numbers) and of the re-formation of other solids.

When water is boiling, for instance, the vapour disappears into the air which becomes warmer. This is the result of the regrouping of the 20 faces of water (icosahedron) into two particles of air (octahedra) and one of fire (tetrahedron). On the other hand, as a consequence of the nature of the triangles involved, the earth cannot be transformed into fire, air or water. Particles of several grades of the element water compose the juices of plants, and their different characters result from the varieties of combinations. Plato mentions several kinds of juice "containing fire", which includes wine.

The immortal soul of man is confined in the "spherical body" of the skull, the rest of the body being considered as a vehicle. The "marrow" is, for Plato, the fundamental life substance in which the immortal soul is located. It fills the skull and the spine and is made of proper proportions of fire, air, water and earth, the same in all animals.

"The god set apart from their several kinds those triangles which, being unwarped and smooth were originally able to produce fire, water, air and earth of the most exact form. Mixing these in due proportion to one another, he made out of them the marrow". (translation by Cornford[21])

The seed which finds a receptacle and an outlet in the sexual organs is a part of the marrow, and continuous with the brain. The "spirited part" of the mortal soul is linked to the marrow by anchor cables. This spirited part is situated in the chest (heart and lung) and the "appetitive part" is in the belly (liver and spleen).

Around the marrow a solid shield of bone was framed. The recipe for the construction of bone is given with great precision and not without a touch of Socratic irony:

"Having sifled out earth that was pure and smooth, he kneaded it and soaked it with marrow, then he plunged the stuff into fire, next dipped it in water, and again in fire and once more in water; but thus shifting it several times from one to the other he made it insoluble by either. Of this then, he made use, first to turn a sphere of bone to surround the creature's brain, and to this sphere he left a narrow outlet; and further, to surround the marrow along the neck and back, he moulded out of bone vertebrae, which he set to serve as pivots, starting from the head through the whole extent of the trunk". (translation by Cornford[21])

Flesh, skin and hair cover and protect the whole. Plato considers that, from the food, in which the god has provided the substances needed to nourish all the parts of the body, like by like, blood is formed in the belly by the action of fire and distributed to the body, which it nourishes, through two veins which may be identified with the hepatic (right) and the splenic

References p. 50

(left) veins. The pumping is ensured by the respiratory apparatus, a mechanical device. The lungs receive breath and drink which provide cooling for the "knot of the veins" (the heart), the principal seat of the fountain of fire. Respiration provides the force which keeps the internal fire in motion and enables the sharp fire tetrahedra to cut up the food, pushing through the veins the blood formed by this process. Blood, the stream of nourishment, contains all substances necessary to replace the waste in organs, resulting from the assaults of the elements outside the body, stimulating the particles to fly off to seek their likes outside. Progressing through the veins, the blood leaves to the organs the various substances composing them. They are attracted like to like by those bodily organs and they repair their waste.

Each substance of the food, which has been reduced to fragments, replenishes that part of the same in the blood which has moved towards its own kind inside the organs. Plato recognizes the existence of several kinds of fire, the sides of their tetrahedra consisting of two or more triangular elements.

The elements in the food have a weaker structure and are more soluble than the young marrow, in which the fire particles are sharper and consequently penetrate into the food and decompose it. This effect on the food becomes more difficult when the triangles of the marrow and consequently their particles are no longer perfectly replicated during their continual reproduction. Owing to this imperfection, they are not as tightly joined as they were previously. The task of decomposing and assimilating the food is therefore more and more difficult, because the process of decomposition predominates over the composition process. Finally death comes as a consequence of disruption of the marrow under the assaults from outside.

How are the triangles damaged through wear? It is said in the *Timaeus* that "the root of the triangle is loosened". Cornford[21] renders this as a bending of the triangle's sides, whereas Lindskog[22] takes it to mean that the points are blunted, a subject on which we shall not linger.

In the *Timaeus*, as well as in the writings of Democritus, the problem of retaining the very small particles of heat (and air) in the body is considered. Streams of heat (and air) pass out through all the pores (not only through the mouth and nostrils) but the continuous loss of heat is compensated. As in the theory of the atomists, heat (fire) and air function constitute a driving force in the vascular system. According to Plato, after fire has decomposed the food as stated above, breath follows on its way through the body.

"Thus filling the veins with substances from the stomach by pouring into them the decomposed nutriments. These decomposed nutriments, taking on the colour of fire, form the blood, which nourishes the entire body in the course of its circulation." (translation by Cornford[21])

The concept of composition and decomposition explains how, in spite of its continuing state of decay, the organism is able to survive.

The *Timaeus* is an admirable poem and must be taken as such, but it must be kept in mind that, as Cornford[21] underlines, "there is no key to poetry or myth".

5. Metabolism in Aristotle's natural philosophy

The philosophical genius of Aristotle[23] did not preserve him from accepting a large number of errors. His biology is greatly influenced by the knowledge of the hippocratic treatises on one hand and of the *Timaeus* on the other. Aristotle rejects the integral mechanism of any presocratic thinkers and he follows the teleological concept of the *Timaeus*, whose origin is probably in Diogenes of Apollonia[24,25]. While Plato viewed life as imposed on matter, Aristotle considered life as being organization, in which he differs from the atomists who were emergentists[19]. But it must be underlined that certain conceptions are common to Plato and Aristotle as well as to the atomists. This is the case for the pre-biochemical concepts of dynamic permanence, formulated by Heraclitus, of the dynamic equilibrium of contraries, introduced by the Pythagoreans, and of the role of vital heat in processing food and converting it into tissues (direct assimilation).

In Aristotle's general system of natural philosophy, matter does not exist by itself. For him, there is an ultimate matter which can take different *forms*, those of the four elements, able themselves to take, in their different combinations, a number of other different *forms*. The ultimate matter is able, through these two levels of *forms* to take an infinity of aspects*.

Aristotle's theory of metabolism is based on the idea that all organic life presupposes heat. As stated by Regnell:

"The process of nutrition which is common to organisms, cannot proceed without heat. It is heat, in fact, which transforms food. All other biological and mental functions depend on

* It may recalled that, in our modern theory, the information carried by the orderly combinations of five nucleotides and of twenty amino acids accounts for the extreme diversity of organisms, which we relate, not to the building blocks but to the order in which they are assembled.

References p. 50

Plate 3. Aristotle.

the process of nutrition. At least this is so in mortal beings, it is added in *De anima*. And the maintenance of dynamic permanence of course requires nourishment. Consequently, Aristotle says, "bodily warmth or fire is a fundamental condition for the maintenance of life."(ref. 2).

While, for the atomists, the material cause of things was their unique cause, Aristotle institutes a distinction between living creatures and lifeless things through the fundamental importance of nutrition, growth and reproduction, the function of the "nutritive soul" or "vegetative soul". (On the philosophical implications of this concept, see Regnell[2] and Hall[19].)

According to Aristotle, the heat of the heart expands the nutrient fluid which continuously flows to the heart. When the lungs cease to work, the fire goes out. The heart does not move by itself. It heats the nutrient fluid and expands it, and this is the cause of the heart's motion. Breathing does not repair the loss of fire but moderates bodily warmth.

The heart is the central organ, the instrument of the soul. The blood comes from it, and from the blood comes *pneuma*. *Pneuma* is an air-like substance which contains vital heat, and Aristotle assumes the dependence of the soul upon the *pneuma*. (On these points see Regnell[2].) From Aristotle's *Meteorologica* it may be concluded that he considers *pneuma* as "a powerful and rapidly moving substance which is the source of movement as well as the source of heat"[2].

The *pneuma* imparts vital heat to the living being. Vital heat "concocts" the nutriment into blood by which the different parts of the organism are renewed and eventually increased while retaining their specific pattern. To explain this fact, Aristotle introduces the idea of the "nutritive soul" which is the "form of the living being"[2].

According to Aristotle, food is broken up in the mouth mechanically and the small pieces undergo "concoction" (*pepsis*) in the stomach. From the intestine, the semi-concocted food is brought to the heart by the mesenteric veins, and a second concoction is accomplished in the liver and spleen as well as in the heart, changing the food into blood. Both concoctions are accomplished by the "vital heat" liberated from the *pneuma* in the heart. A third conversion takes place in the assimilation of blood into tissues, under the action of the active principle of growth, the nutritive soul. The nutritive soul converts the food that is potentially flesh into flesh, with the help of cold, and the food that is potentially bone and sinew into bone and sinew with the help of heat. (For individual references to these aspects in Aristotle's writings, see Hall[19].)

References p. 50

6. Galen's humouralism and his views on metabolism[26]

Galen's doctrine of *pneuma* (spirit)[19,27,28] is expressed in his concepts of animal spirits and vital spirit, playing a determining role in his views on metabolism and respiration. Galen (A.D. 130–200) considers that, during the metabolic process, exchanges of *qualities* and not of substances take place. It is from Aristotle's teachings that the idea according to which the four elements consist of exchangeable qualities is derived. The stoic philosopher Chrysippus writes:

"Passive and motionless matter is the substrate of the qualities, while the qualities are *pneumata*, aerial tensions, inherent in the parts of matter and determining their form". (translation by Sambursky[29])

For Galen the four elements consist of combinations of the four qualities: hot, cold, dry, moist (*primary qualities*). Fire is hot and dry, water is cold and moist, air is hot and moist, earth is dry and cold. Density, lightness, hardness, etc. are *secondary qualities*.

Following Aristotle and the stoic school, Galen believed that an element can be transmuted into another by changing the proportions of its primary qualities. The organism, in Galen's system, is a combination of primary qualities, modified by secondary qualities. Metabolism, for him, begins with the transformation of food into *humours*, composed of the same four elements found in food. By a transposition of the qualities of food under the action of the heat of digestion, food becomes lymph, lymph becomes humours, humours supply the material for the body tissues. According to this view, the *qualities* (non-material) are the only building blocks of the organism, superimposed on the universal carrier, the hypothetical *ousia*.

The stoics, as well as Aristotle, had rejected the theory of fire as proposed by the atomists who considered it as a movement of the atoms. Aristotle's concept was that "fire is setting free more of the heat bound to the inflammable substrate by increasing the motion of the invisible heat"[30]. This was in agreement with the old idea that combustible substances give up their bound heat, air being necessary because it furnishes a continuous supply of heat (from *pneuma*), replacing the liberated, previously bound, heat.

Experimentally, Galen found that vital combustion does not consume measurable amounts of air (owing to his ignorance of the replacement of O_2 by CO_2 in the air used), and he concluded that, as in a burning flame, only a *quality* of air is absorbed in respiration and not a material component. The vital factor from the air is, according to Galen, absorbed by the body

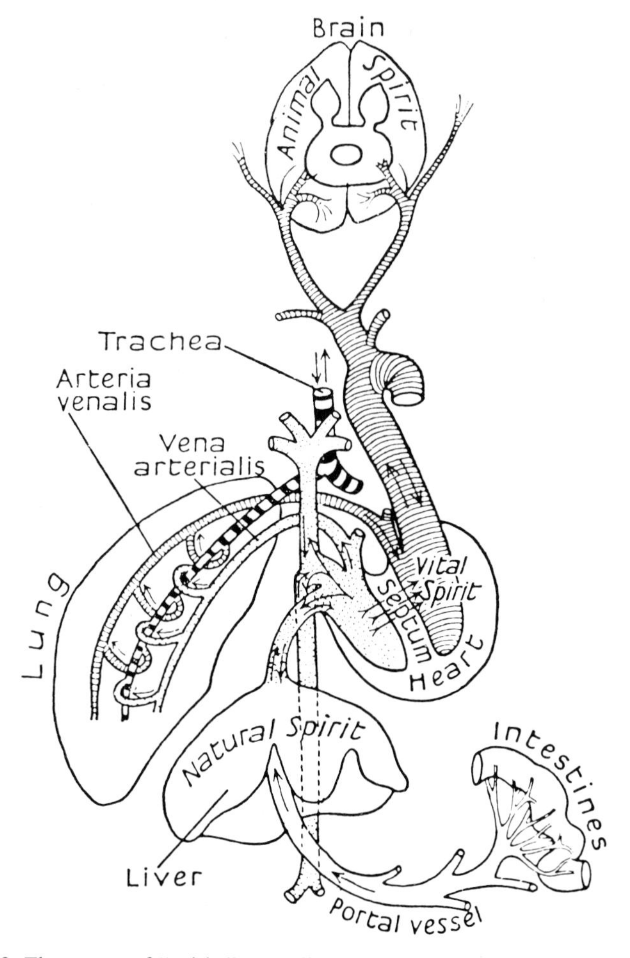

Fig. 2. The course of "spirits" according to Galen. Only one lung is shown.

at three different sites: the lungs in adult animals, the placental vessels in the foetus, and the water bathing the gills of fishes. At the level of the lungs, he believed that blood entering into contact with air was enriched not in a substance but in *pneuma zotikon*, the vital spirit (a quality invisible, not measurable). For Galen, as for the early Greek philosophers, the combustive process took place in the left ventricle of the heart.

The food taken into the mouth is first modified to some extent by the action of saliva (phlegm). More food is changed in the stomach by the

References p. 50

action of phlegm, bile, pneuma and vital innate heat. A first elaboration results, accompanied by an elimination of the useless parts, while the useful parts penetrate into the intestine and from there to the veins, which lead them to the liver, where a second elaboration takes place which converts the food into blood.

We see that according to Galen, contrary to the atomists who thought that food was merely broken up into its constituents and absorbed unaltered, the food undergoes a real qualitative alteration during the nutritive process, becoming similar to the substance of the organism. The Galenic conception of circulation has been the subject of error for many authors. For example, it has been claimed that there was a reflux from the right ventricle and a discharge of waste through the pulmonary artery. Furthermore, it has been generally assumed that Galen thought that blood was "distributed by the liver throughout the venous system which arises from it, ebbing and flowing in the veins"[31]. That this often-mentioned "ebb and flow" is not part of Galen's views is illustrated by the interpretation by Fleming of Galen's text on circulation:

"Blood on entering the right ventricle must pass by a one-way valve opening inward, so that only an insignificant portion can elapse into the vena cava whence it came. Some of the blood passes directly from right to left through the interventricular septum. But much, and apparently most of the blood moves into the arterial vein (and pulmonary artery) past a one-way valve opening outward from the ventricle. On contraction of the thorax, the blood in the arterial vein, its retreat cut from behind can only go forward into the arterial system of the lungs (in modern usage, venous). Whether Galen's venous artery, corresponding to our pulmonary veins, then carries blood to the left ventricles is in question. He almost certainly thinks of the venous artery as conveying the inspired air, in some form or other,—or at least some quality derived from the air,—from the lungs to the left ventricle. In the opposite direction smoky wastes are undoubtedly borne from the ventricle to the lungs by way of the venous artery. This process is made possible in Galen's view by the comparative insufficiency of the mitral valve opening in the heart. The blood in the left ventricle passes into the aorta through an aperture guarded by a one-way valve opening outward." (ref. 32)

As stated by Wiberg[33], for Galen,

"the heart was *an organ for the Respiration* or, properly speaking, such was the function of the left half of the heart; the physiological function of the right half of the organ was to produce nourishing blood for the lung and materials for the formation of *pneuma* in the left ventricle while the role of this ventricle was to get the air that was necessary for the formation of *pneuma*, and this *pneuma* was indispensable for the respiration of the tissues." (ref. 33)

The blood formed in the liver is "on the way" to becoming flesh, which is the highest form of assimilation of nutrients. Blood undergoes different degrees of coction and varies in thickness, heat, etc. Each type of blood nourishes the organ which is most like it. The four humours are also derived from different types of blood. A proper mixing of humours is the condition of health, improper mixing leading to disease. The purest form of blood is the variety charged with *pneuma*. The inhaled air serves to cool the excess heat of the heart and is breathed out again, the immature *pneuma* passing through the heart and mixing with the blood for the production of the essential innate heat. The *nutritive faculty* is responsible for the assimilation of nutrients, nutrition resulting from an attraction or assimilation of that which nourishes to that receiving nourishment. Food is brought by the veins to the parts that are being nourished.

The process of assimilation occurs in steps:

"How would blood ever turn into bone without having first become, as far as possible, thickened and white? And how could bread turn into blood without having gradually parted with its whiteness and acquired redness? Thus it is quite easy for blood to become flesh, for if Nature thickens it to such an extent that it acquires a certain consistency and ceases to be fluid, it thus becomes original newly formed flesh, but in order that blood may turn into bone much time is needed and much elaboration and transformation of the blood. Further it is quite clear that bread, and more particularly lettuce, beet and the like, require a great deal of alteration to become blood." (translation of J. A. Brock, cited by Leicester[5])

Many organs are needed to alter food. Each organ has its attractive faculty: kidney for urine, stomach for nutrients, gall bladder for yellow bile, etc. Galen chose the elements of his system from the wide variety of speculations around the ideas of *innate heat* (fire), *pneuma* (air) and *coction* as the combining power. He held them together with basic principles of teleology (this appealed strongly to his Christian and Muslim successors). The humoural doctrine of Galen played a very important role in the history of medicine. It also brought for the first time a hierarchy of material components of the organism and it must be considered as a remote ancestor of the concepts of "indirect nutrition" and of the "internal environment" (see Chapter 10).

Hall[19] has endeavoured to derive, from the variety of contexts in which Galen has written about humours, a unified scheme to which the reader is referred.

7. Retrospect

In the biological system of Aristotle which was to continue until the 18th century, we find an endeavour in which most human obstacles to the scientific spirit are embodied. To introduce the concept of unity of the

References p. 50

universe, Aristotle combined in his proto-biochemical theories the concepts of matter-of-life and of forces-of-life into the system of the four elements and of their qualities. To this he added the concept of the purposefulness of nature and that of final cause. While the conceptions of the atomists were mechanistic, those of Plato and Aristotle introduced the curses of teleology and vitalism. Nevertheless, as well as Heraclitus, Plato and Aristotle believed in the dynamic permanence; and like the Pythagoreans, they believed in the dynamic equilibrium of opposites. Atomists, Plato and Aristotle all believed in the role of fire (innate heat, vital heat) in the processing of food and in its conversion into tissues (assimilation).

It has been very brilliantly claimed by Bachelard in a number of delightful books[34] that the doctrine of the elements was rooted in the archetypes of the human categories and is even more dependent on the discursive activity of the human mind than on the result of sensorial activities.

The concept of the elements, a manifestation of the tendency to substantiation, was for centuries an epistemological obstacle to the development of the concept of pure substance. The elements, corresponding in the system of the Greek philosophers to our concept of matter, represented, combined with the concept of *form*, the solution to the proto-biochemical problem of the opposition of permanence to continual change. This is still clearly expressed in a sentence written by Charles Bonnet in the 18th century:

"De l'invariabilité des espèces au milieu du mouvement perpétuel qui règne dans l'univers, se déduit l'indivisibilité des premiers principes des corps." (ref. 35)

With the concepts of dynamic permanence, of dynamic equilibrium of opposites, and of form, solutions of indured laziness of mind were always at hand to explain all forms of material changes in living organisms.

The Greek proto-biochemists, the Hippocratics included, showed the tendency, appearing as well in the other chapters of Greek science, to observation, speculation and formulation of deductive theories. The Hippocratics observed the sick, whereas Empedocles, Anaxagoras, Aristotle or Plato observed nature, but the process of reasoning was the same in all cases, leading to qualitative, teleological, macrocosm–microcosm deductive systems.

The Greeks did, nevertheless, define a number of themes which were transferred from proto-biochemistry to biochemistry. Among these themes are the concept of composition and decomposition (waste and repair, catabolism and anabolism) ensuring the dynamic permanence of the organism; the concept of direct assimilation, according to which repair is ensured by the introduction of constituents of the food similar, or rendered similar, to the worn-out parts they replace and the concept of the balance of opposites still extant today in enzymology, in endocrinology and even in molecular biology (operators and repressors in bacteria). To these long-lived components of ancient Greek thematics must be added the idea of the *go* of life conceived as vital heat, depending upon the *pneuma*, a relation that has gone through a number of expressions. It must be recognized that, in our own theories, the concepts of the energy of oxido-reductions and of oxygen as a partner in biological oxidations, though different in their content, occupy similar positions in the frame of the theme of dynamic permanence as did vital heat and *pneuma* in the conceptual formulation of the ancient Greek theories.

References p. 50

REFERENCES

1 J. Diels, *Die Fragmente der Vorsokratiker*, 7th ed., edited with additions by W. Kranz (a photographic reprint of the 5th ed.), Berlin, 1954.
2 H. Regnell, *Ancient Views of the Nature of Life*, Lund, 1967.
3 W. A. Heidel, *Antecedents of Greek Corpuscular Theories*, Harvard University, 1911.
4 C. Lejewski, in E. McMunn (Ed.), *The Concept of Matter*, Notre Dame, Ind., 1963.
5 H. M. Leicester, *Chymia*, 7 (1961) 9.
6 F. J. Collingwood, *Philosophy of Nature*, Englewood Cliffs, 1961.
7 J. Burnet, *Early Greek Philosophy*, 3rd ed., London, 1930.
8 E. Zeller, *Outlines of the History of Greek Philosophy*, London, 1931.
9 F. M. Cornford, *The Unwritten Philosophy and Other Essays*, Cambridge, 1930.
10 J. Bollack, *Empedocle*, 4 vols., Paris, 1965.
11 H. Erhard, *Arch. Gesch. Med. Naturw.*, 35 (1965) 117.
12 W. H. S. Jones and E. T. Withington (Eds.), *Hippocrates and the Fragments of Heraclitus*, 4 vols., London and Cambridge, Mass., 1922–1931 (Loeb Classical Library).
13 R. Joly, *Hippocrate, Médecine Grecque*, Paris, 1964.
14 R. Joly, *Le Niveau de la Science Hippocratique*, Paris, 1966.
15 R. Joly, *Recherches sur le Traité Pseudo-Hippocratique "Du Régime"*, Paris, 1960.
16 C. E. A. Winslow and R. R. Bellinger, *Bull. Hist. Med.*, 17 (1945) 127.
17 C. Bailey, *Epicurus, The Extant Remains*, Oxford, 1926.
18 Lucretius, *De Natura Rerum*, translated by W. H. D. Rouse, London, New York, 1924.
19 T. S. Hall, *Ideas of Life and Matter*, 2 vols., Chicago and London, 1969.
20 C. Bailey, *The Greek Atomists and Epicurus*, Oxford, 1928.
21 F. M. Cornford, *Plato's Cosmology, The Timaeus of Plato Translated with a Running Commentary*, London, 1937.
22 C. Lindskog, *Platon*, Vols. I–VI, Stockholm, 1920–1926.
23 Aristotle's works (Loeb Classical Library), London and Cambridge, Mass. (Several editors and dates.)
24 W. Theiler, *Zur Geschichte der teleologischen Naturbetrachtung bis auf Aristoteles*, Zurich, 1924.
25 R. Joly, *Rev. Philosoph.*, 158 (1968) 219.
26 Galen's works in K. G. Kuhn, *Medicorum Graecorum Opera Quae Extant*, Leipzig, 1821–1833, 20 vols.
27 L. G. Wilson, *Bull. Hist. Med.*, 33 (1959) 293.
28 O. Temkin, *Gesnerus*, 8 (1950) 180.
29 S. Sambursky, *The Physical World of the Greeks*, London, 1956.
30 R. E. Siegel, *Galen's System of Physiology and Medicine*, Basel, 1968.
31 Ch. Singer, *A Short History of Anatomy and Physiology from the Greeks to Harvey*, New York, 1957.
32 D. Fleming, *Isis*, 46 (1955) 14.
33 J. Wiberg, *Janus*, 41 (1937) 237, 244.
34 G. Bachelard, *La Psychanalyse du Feu*, Paris, 1938; *L'Eau et les Rêves, Essai sur l'Imagination de la Matière*, Paris, 1942; *L'Air et les Songes, Essai sur l'Imagination du Mouvement*, Paris, 1943; *La Terre et les Rêveries de la Volonté, Essai sur l'Imagination des Forces*, Paris, 1948; *La Terre et les Rêveries du Repos, Essai sur l'Imagination de l'Intimité*, Paris, 1948; *La Poétique de l'Espace*, Paris, 1957; *La Poétique de la Rêverie*, Paris, 1961; *La Flamme d'une Chandelle*, Paris, 1961. This series of books devoted to the psychoanalysis of imagination follows, in the work of Bachelard the series of studies on scientific epistemology (1928–1938) which constitutes his major contribution and which culminates in his classic: *La Formation de l'Esprit Scientifique*, Paris, 1947.
35 Ch. Bonnet, *Contemplation de la Nature*, 2 vols., Amsterdam, 1770.

Chapter 2

From Alchemy to Iatrochemistry

1. Postgalenic proto-biochemistry

The heritage of ancient Greek proto-biochemistry was preserved during the following periods mainly in the form of the theory of the four elements. This theory applies, of course, to the whole of the Cosmos but it would be an error to consider, as is sometimes done, that it did not individualize the living from the rest of the Cosmos.

Made of the same four elements associated in particular proportions, the organisms were considered as showing an autonomy depending on composition and decomposition in a dynamic equilibrium, and even the differences of composition of their parts was clearly stated in the context of "assimilation"*.

Histories of chemistry relate the history of the very ancient part of practical chemistry of which the whole archaeology of knowledge relates the experiments with glass, metal, cements. Technical treatises on what we would call chemical matters have existed since antiquity in the form of recipe books. But in these treatises such a concept as metal being a distinct elementary substance with characteristic properties was never attained by the Ancients.

As we saw in Chapter 1, Aristotle admitted the existence of an innate heat, necessary to the metabolism and functions, and cooled by air. His *pneuma* was innate.

In a more properly defined proto-biochemical context, the pneumatology of Galen, revived from Arabic sources, became dominant in the Western

* Selection of different constituents of the food by different parts according to the composition of their parts (Anaxagoras); their concoction by the nutritive soul converts what is potentially a part into that particular part (Aristotle); each type of blood made in the liver nourishing the organ that corresponds to its nutritive faculty (Galen).

References p. 79

Plate 4. Jean Fernel.

world. As we have seen, he considered the spirits (*pneumata*) as being derived from the air, the material from which the *pneuma* originates.

The four elements and the *pneuma*, in the meaning proposed by Galen, remained incorporated in Western thought for centuries.

In an encyclopedia of the knowledge of the time, *Li livres du Trésor*, which had a great popularity in the course of the 13th century, by Brunetto Latino, we find a chapter on the nature of things in which it is explained that the elements are light (air, fire) or heavy (earth, water), each one having extremities and middle sections. In flame for instance, the upper part is lighter than the lower one. The four qualities also have gradations. Elements and qualities are mixed in variable proportions. The more subtle particles of which animals are made have been used to form the birds. This explains why the bird flies. If an eagle flies better than a duck, it is because he contains a greater proportion of subtle parts. Animals are kept alive by four virtues: the appetitive, the retensive, the digestive and the expulsive. These virtues correspond to the four elements. The fire, warm and dry, induces eating and drinking (appetitive virtue). The presence of the four elements is necessary for the establishment of the vital equilibrium[1].

Agrippa von Nettesheim (1486–1535) teaches that,

> "stones are made of earth and heavy, that metals are watery and fusible, that in the plants the air dominates, as the fiery vital force dominates in animals". (translated by author after the German text in Rotschuh[2])

As stated in Chapter 1, Galen and subsequent doctors of "Galenic medicine" considered metabolism as a series of "digestions", engendered by heat. Through heat, the nutritious portions of food and drink were successively converted into chyle, into the four humours and finally assimilated. To epitomize the concept of metabolism of the "schools", it finally suffices to recall that they believed in a "coction" of the food in the stomach under the influence of the "innate heat", followed by "fermentation" in the liver, and by a third digestion due to bilious "acidity".

Fernel (1497–1558) (bibliography of his writings in Sherrington[3]), one of the most prominent Galenists, attempted to improve on Galen in the domain of metabolism, by considering the analogy of the coction of the stomach with the cooking of food, concluding that the former involves an "innate spirit" as well as a "peculiar natural heat". He also considers the analogy of the blood-making process, thought to take place in the liver by Galen, with fermentation of wine. He recognizes that in both processes

References p. 79

Plate 5. Oratory and laboratory. The alchemist in prayer.

deposits appear. In the liver, the deposits are the black bile, stored in the spleen, and the yellow bile, stored in the gall bladder.

In his book *De Natura* (1565)[4], Bernardino Telesio, although strongly opposed to an overreliance on the authors of antiquity and just as strongly in favour of an observational approach to nature, remained a Galenist in respect to some of his most fundamental physiological views. In *De Natura* he gives emphasis to the living organisms. His metabolic system incorporates the notion that the stomach "cooks" the food for the liver to make blood, but that the motion of the oesophagus and of the stomach represent the rejection of portions of the food improper to its use by the liver. For Telesio, hunger and thirst derive from the pouring of acid juice from the spleen into the stomach. When the stomach is empty, the acid reaches the fibres of the wall, causing hunger and a movement which helps in breaking up the food. When it reaches the liver, the chyle still contains impurities. A black sour bile is drawn up by the spleen while a yellow bitter bile goes to the gall bladder. Another excretion, opposed by its thinness to the pernicious thick biles, passes off as urine.

2. Convergences in alchemies

While their predecessors were limited to discursive schemes of nature based on the four elements and the Aristotelian or the Galenic *pneuma*, the alchemists were the first proto-biochemists who used a laboratory. This laboratory, as shown in graphic representations, was also an oratory (see Plate 5). As they considered the whole universe as being alive, and were the first to perform a chemical analysis of living matter, the alchemists were essentially proto-biochemists. But they did not, of course, reason according to the rational ways of the scientific method developed by Galileo. Their conceptual system consisted in one variety or another of religious or philosophic theory.

Under what we consider now as alchemy, we must recognize a number of different varieties of polyphyletic origin, among which it is difficult to describe the variations and interrelations as well as the priorities, owing to the uncertainty about the chronology of the extant manuscripts. Starting from a philosophical and religious system to study the whole world, which they considered as alive, the varieties of alchemy, either Chinese, Greek, Arabic, Indian or European, converged in the use of a number of types of apparatus. These, though different in their construction according to place

References p. 79

and time, and derived from cooking technology, including treatment of fermented beverages, were used for sublimation, distillation, etc. As noted by Bachelard[5], the importance of distillation was very great in prescientific periods. It gave to the unconscious mind of the researchers a picture of their transmigration dreams and the still was, so to speak, considered as a diminutive of the universe, this vast alembic whose still-head is like the head of the human microcosm and whose head-beak is similar to the nerves adapted to the brain.

3. Chinese alchemy

While the outline of the ancient Greek philosophy developed in the frame of the cosmic system, another entirely different system had been conceived in China. This dynamic system applied a philosophical framework to the changes of matter and endeavoured to interpret nature in terms of life.

Chinese alchemy has its origins in Taoist religious beliefs. Sivin[6] traces its known historical steps as follows. During the 8th century B.C. one can trace in Chinese thought a belief in the possibility of physical immortality. In the 4th century B.C., immortality was thought to be obtainable by drugs as well as by a number of other methods. Later the idea appears that such drugs must be made by man and not naturally obtained. In the extant sources, the transformation of cinnabar into gold does not appear before 133 B.C. It is clearly associated with the cult of immortality, but indirectly, for instance by eating from utensils made of that kind of gold. Later the notion appears according to which some potable form of natural or artificial gold (as well as other drugs) could bring about "transcendence". Artificial gold (not a form of currency, but an imperishable substance in the Chinese context) was emphasized by Chinese alchemists as conferring longevity or immortality.

> "When the golden powder enters the five entrails,
> A fog is dispelled, like rain-clouds scattered by wind
> Fragrant exhalations pervade the four limbs;
> The countenance beams with well-being and joy..." (translation by Waley[7])

It is only in the mid-second century A.D. that theoretical elements can be found, but they are of so complicated a nature that it appears that some antecedents have been lost.

Sivin[6] has found in the Chinese alchemical literature approximately a thousand titles of *elixirs* of immortality, but he estimates that the number should be reduced, owing to the use of synonyms, to about half. The laboratory equipment of the Chinese alchemists, and particularly their stills, has been described by Ho Ping-Yü and Needham[8].

4. Greek alchemy

In Greece, alchemy started in the first century A.D. and continued through the Byzantine culture. Though it lacked the "macrobiotic" component, this form of natural philosophy can be called alchemy since it applied a philosophical framework to the changes of matter and interpreted nature in terms of life. The philosophical tendency that was most influential on this system was the stoic one. Of particular importance to alchemy was the stoic concept that animals and plants reproduce their kinds according to fixed types by virtue of the *pneuma*, or spirit, operating in each kind in a specific fashion. The principle by which the *pneuma* worked in an individual was termed the *spermatikos logos* or "seminal formula"; it implied that reproduction of a given type took place from a "seed" through the impelling force of the *logos*, which was regarded as a rational organizing principle founded in the Divine Intelligence. To extend this to the mineral world as did the alchemists, was thus a simple step.

For the stoics, *pneuma* originated as part of God who was Divine Fire (undoubtedly inspired by the ideas of Heraclitus). When part of the Divine Fire condensed and coarsened, the Divine Nature was lost and it became the "elements", air, water and earth, out of which the remaining Divine Fire (as "Active Reason") fashioned the Cosmos. At the centre was the world, partially covered by an envelope of Divine Fire from which God interpenetrated the whole of matter in the form of *pneuma*. Eventually the Cosmos would undergo destruction by the Divine Fire but would re-emerge in stages from a remaining "seed" through the action of *pneuma*[9].

Much material concerning this philosophical background of Greek alchemy has been provided by a number of recent studies[10,11].

The alchemists looked for the quintessential spirit, which could bring seeds of noble metals, present in small amounts in base metals, to grow as if alive. As it was believed that metals such as gold developed from a seed like a plant, it remained only to isolate these seeds of gold and to select the proper earth for their development. In other varieties it was the spirit (in the form of the philosophers' stone or a substance of which small quantities will transmute a large quantity of base metal into precious metal) which multiplied itself.

References p. 79

Even before the "macrobiotic" element was introduced into western alchemy, it had recourse to elixirs obtained by distillation to perform the transmutation (white elixir and yellow elixir, respectively considered to convert base metals into silver or gold).

When the alchemists endeavoured, for instance, to change copper into gold, they tried to remove the *form* (as defined by Aristotle) of copper (or symbolically to accomplish its death and corruption) and to introduce a new *form*, that of gold (symbolically a resurrection). A subtle matter, the *pneuma* of the Stoics or the spirit (which is not to be confused with the soul), could in the theoretically possible transmutation of any matter to any other, either become a matter (for instance a metal) or help to control its birth and produce new *forms*. A generation of new *forms* could only obtain after the corruption of the matter to be transformed.

5. Arabic alchemy

The Arabic culture inherited the treasure of knowledge accumulated by the ancient Greeks, including alchemy. The Nestorians and Monophysites, when they were exiled from Byzantium, carried it to Persia and Syria, and after the rise of Islam the relevant treatises were translated into Arabic. On the other hand, a knowledge of Chinese alchemy and of its "macrobiotic" component was transferred by Chinese travellers to Bagdad, the Arab capital, around the 9th century. Arabic alchemy was not simply a conservatory of ancient knowledge, but it combined the transmutation and the macrobiotic components. It was most active from the 9th to the 11th centuries.

The Jabirian collection of alchemical treatises which used to be attributed to a legendary author by the name of Jabir, but is now considered as emanating from a group of natural philosophers known as "Ihwan al-Safa" (Brethren of sincerity), divides the known substances into spirits (volatile bodies such as camphor or sulphur), and metals and bodies (non-volatile and non-metallic). Metals, as in the Greek alchemy, were considered to be a combination of body and soul (spirit).

A theory which is expounded in the Jabirian treatises* is that the metals are composed of "mercury" and "sulphur". This is derived from Aristotle (*Meteorologica*, Book III, Chapter 6, p. 378c), who supposes that the metals are made of two "vapours" (or "exhalations"), one dry (smoke)

* The following text concerning the Jabirian treatises is derived from Taylor's[12] analysis.

and one moist, rising through the earth. The flammable part of metals is considered as "sulphur".

The Jabirian treatises nevertheless consider that the metals are ultimately composed of the four elements and present the four qualities, one externally and one internally, as follows:

	Outer qualities	*Inner qualities*
Gold	hot-moist	cold-dry
Silver	cold-dry	hot-moist

A transmutation was possible through an alteration of these qualities, and the process had been learned by the Arab alchemists from Galen, whose therapy was based on a process of rectification of the proper equilibrium of the qualities. To cure the baseness of the metals, the Jabirian alchemists used a medicine, the "supreme elixir". They considered that it would be possible to find out the proportions of the four elements in each body and to convert it into another body by adjusting the proportions of the elements through adding pure elements mixed in the right proportion in the form of an "elixir".

Of course the metals could not be analyzed into anything comparable to the four elements, but this consideration is not applied to the organic bodies which could be analyzed by distillation. This idea comes from the Greek alchemists (as we shall see, they distilled eggs). The Jabirian alchemists distilled a number of plant and animal products. They obtained "sal ammoniac" (ammonia) by distilling dried animal dung. The distillation of organic substances provided the alchemist with supposed elements, each with two qualities:

"*(1)* A liquid, which was the element of *water* (cold and moist),

(2) Something he called oil or grease, which was an inflammable body which he identifies with the element of *air* (hot and moist): this was presumably a mixture of volatile combustible organic liquids and gases,

(3) A combustible colored substance called fire or tincture (probably a tarry body) which he identifies with *fire* (hot and dry),

(4) A dry mineral residue, mainly charcoal, which he identifies with the element of *earth* (cold and dry)". (Taylor[12])

Each supposed element having two qualities, the purpose is to isolate pure elements, with only one quality, in order to add, for instance, coldness to a metal without adding moisture or dryness.

References p. 79

In order to transform water (cold and moist) into what is only cold and no longer moist, he distills and redistills water, adding "very dry" substances to remove the moist quality. The Jabirian treatises teach that after hundreds of distillations, the water, now deprived of its moist qualities, solidifies as salt. It may be recalled that we must wait until Lavoisier's experiments to disprove the transformation of water into earth by repeated distillations.

The alchemist could obtain not only the element which is wholly cold from the "water" resulting from the distillation of organic matter, but he could also obtain the moist from his "oil", and the dry from his "earth". From his "tincture" (hot and dry) he could obtain the element which is wholly hot, and which seems to have been the precursor of what was later called the "philosophers' stone", composed of what is lacking in base metals and present in gold.

"Having obtained the "pure elements", the alchemist was required to mix them in specific numerical proportions, so forming a suitable "elixir" which was to be applied to the metal by a somewhat complicated process. Transmutation should then take place." (Taylor[12])

The definition of the numerical proportions of the components of the mixture is most interesting. The Jabirian alchemist assigns, through careful weighing out of quantities, a value to each substance.

"If gold is worth 1, the elixir is worth 5. The power of each operation is denoted by a fraction. A sublimation is worth 1/50 and a fusion 1/200. The conclusion reached by such a calculation is written as follows:

$$(\text{Gold}) \times (\text{Fusion}) \times 1000 = (\text{Elixir})$$
$$1 \times \tfrac{1}{200} \times 1000 = 5$$

The conclusion is that 1000 fusions should convert gold into elixir." (Taylor[12])

The authors of the Jabirian corpus were contemporary with Razes, the famous physician and alchemist Al-Razī (860–925). We may learn another, and similar version of his alchemy from a translation of his *Book Secret of Secrets*[13].

Heym epitomizes as follows the system of concepts on which the experiments of Al-Razī are based:

"The purpose of all the experimental work is the improving of base metals, the transformation of lead, tin, copper and iron into silver or gold, and the improving of ordinary stones, such as pebbles, glass, and ordinary crystals into red Yaqut, green emeralds and other precious stones.

"The means by which these transformations are to be accomplished is a powder or a liquid, produced according to a laborious method of work. This powder or liquid has an effect comparable to a strong medicine or a virulent poison when it permeates the base metals and the pulverized stones, and transforms the whole substance into silver or gold, or into precious stones.

"The possibility of obtaining a species of matter which has these miraculous properties is based on the theory that all forms of matter contain within them a number of specific attributes, which are capable of being increased in potency to the highest degree of effectiveness, or on the other hand, capable of being weakened or destroyed. The method by which this is accomplished consists, on the one hand, of adding other forms of matter with similar or opposite specific attributes to the substances with which one is experimenting or, on the other hand, it consists of certain processes which are described by Al-Razī in his book and which are difficult of comprehension." (Heym[14])

In his book Al-Razī describes the distillation of hair, urine and bat's guano.

6. Mediaeval European alchemy

The Western tradition of full-fledged alchemy was carried to Europe in the 12th century, through the translations from Arabic to Latin first made in Spain from 1130 onwards.

"It mingled with the purely chemical tradition of Alexandria to make up the body of information and practice, knowledge and speculation, which was the alchemy and chemistry of medieval Europe." (Davis[15])

The proto-chemists of Alexandria in the 12th century were not involved in elixir making or in seeking the philosophers' stone. They were imitating gold, purple, silver and precious stones by making alloys or by dyeing, or veneering with layers of oxides, etc. The views of the European alchemists of the 13th century are recorded in the works of Geber[16], probably a Spanish alchemist influenced by Arabic alchemy, who has been in the past confused with the legendary Jabir.

7. Experimental and non-experimental components of alchemy

Nothing is more unhistorical than to consider the alchemists of mediaeval Europe as fantastic humbugs. Their attitude was the first fully proto-biochemical approach to the view that all Nature is alive. This approach was accomplished with a view to the perfection of matter and to the reaching of "transcendence", and later in "a more quasi-religious conception of the regeneration of the soul, otherwise regarded as a spiritual perfection"[9].

Mutatis mutandis, the intellectual attitude of the alchemists, in a very different conceptualist system, was much akin to that of our contemporaries, the "molecular biologists", whose brilliant achievements will be considered in later parts of this History. The molecular biologists decided to find the molecular basis of heredity, boldly looked for it and reached it. The al-

References p. 79

Plate 6. The adept and his wife in prayer under the Philosophers' Egg.

chemists lived in a world not yet provided with the concept of molecules, and in which everything was a creation of a quintessential spirit which they endeavoured to isolate. To that purpose they transformed the equipment of the kitchen into the first chemical laboratories that ever existed, and practised distillation with a few substances to separate the quintessential spirit responsible for all forms of life which they considered as obtaining not only in plants and animals, but also in minerals.

The alchemists, in their cosmic animism, started from life to interpret the universe through a combination of "chemical" experiments and of philosophical or religious doctrines. This combination sometimes led to a very elaborate symbolism*.

In the pre-biochemical emphasis which is put here on alchemy, it may be useful for the reader to consider the ways in which the alchemists, combining laboratory work and oratory, turned their philosophical or religious symbolism into the reality of the chemical operations.

Into the body of their stills, the alchemists introduced what they called "sulphur", *i.e.* a matter, organic or inorganic, which, on heating, produced what could be sublimated or distilled.

In order to obtain all kinds of elixirs, all kinds of living matter were used, among them, for instance, the hen's egg. Taylor[12] describes a recipe of the Greek alchemist Zosimus, which starts from the distillation of eggs, the distilled liquid being obtained in three fractions: first, a clear distillate called "rain water"; then a pale golden liquid called "oil of radish"; then a dark yellowish green liquid called "castor-oil". Now, adds Taylor[12],

"if actual eggs are distilled, we obtain, first, a large quantity of clear liquid, faintly alkaline; then a golden yellow somewhat oily-looking distillate containing ammonium sulphide, ammonia and pyridine bases; lastly, a very dark yellowish thick liquid containing pyridine bases and tarry products. This corresponds quite closely to the alchemist's description of the products and to what he tells us they should do. Thus the second distillate renders arsenic yellow, as he informs us presumably by reason of the sulphites it contains.

"Why should the alchemists wish to distill eggs?

"I suppose that they were seeking to extract the "breath of life", the *pneuma*, that was present in the egg, which of all things has the most obvious potency of generation. Moreover, the yolk of the egg had a promising golden color, and as we have seen, the sulphurous liquids obtained by distilling eggs had a yellow color and could confer it on certain materials of the Art, such as white arsenic and silver." (Taylor[12])

In other words, the second distillate was a "yellow elixir", comparable with the yellow calcium polysulphide solution obtained by boiling lime, sulphur and water and with all yellow liquors used for superficial colouring.

* See Appendix 2, p. 327.

Plate 7. The alchemist wielding the fiery sword above the Philosophers' Egg.

As noted above, it was not until the 12th century that European alchemy knew anything else, in the way of elixirs, than those attacking or colouring metals.

One of the most popular encyclopedias of alchemy, *Splendor Solis*[17] (1582), shows how the alchemists of the time understood the composition of the hen's egg.

"The philosophers take for example an Egg, for in this the four elements are joined together. The first or the shell is Earth, and the White is Water, but the skin between the shell and the White is Air, and separates the Earth from the Water; the Yolk is Fire, and it too is envelopped in a subtle air, which is more warm and subtle, as it is nearer to the Fire, and separates the Fire from the Water. In the middle of the Yolk there is a *Fifth Element*, out of which the young chicken bursts and grows. Thus we see in an egg all the elements combined with matter to form a source of perfect nature, just so as it is necessary in this noble art." (Sheppard[18])

In order to reproduce *in vitro* the process of a rebirth of matter by the addition of an element to the heated metals, a vase was employed, with a symbolic form (*vas mirabile*), a *hermetic vase* in the shape of an egg. This vase became the Philosophers' Egg, symbol of the primordial chaos out of which the living world emerged. The alchemists, who knew about the necessary pairing of rooster and hen, and expressed it by the presence of the *fifth element* in the egg, depicted the Egg or vase as a symbol of the union of opposites, sulphur (King) and mercury (Queen) whose conjunction was considered as a marriage taking place in the Egg considered as a palace,

"constructed by the hands of many craftsmen—the king and his spouse must be quite naked when they are joined together—that their seed may not be spoiled by being mixed with any foreign matter." (Waite[19])

In the 16th century, the orientation of alchemy turns to setting out cloudy schemes of the universe. It develops a vast edifice of symbolism in which gnostic symbolism is adapted to the religious need of the Christian redemption of man. In a plate of Michael Maier's *Atalanta Fugiens*[20] we see the alchemist wielding a sword above the Philosophers' Egg (Plate 7).

Alchemy and gnosticism have drawn on the same stoic philosophy and its belief in

"the unity of the world, deriving from the initial emission of the Divine *pneuma*. All things are One, they arose from the One and will return to the One..." (Sheppard[9])

As we noted above, the Cosmos, surrounded by Divine Fire, could be destroyed by it, but through the action of *pneuma*, could now re-emerge from the remaining "seeds". In the plate from the *Atalanta Fugiens* the flaming (fiery) sword wielded by the alchemist above the Philosophers' Egg

References p. 79

Plate 8. The Chemical Garden.

will perform the ritual slaying to bring about its spiritualization, or transformation into *pneuma*, in order that the chick (philosophers' stone) may arise to conquer iron and fire. It symbolizes the production of the stone (chick) as the result of the application of heat (the flaming sword, the fiery sword). The symbolism of the egg is ever-present throughout the work of the painter Hieronymus Bosch.

This will give sufficient illustration of the simultaneous presence, in alchemy, of natural realities and of their insertion into an intricate web of human symbols marked, as Jung[21] has stressed, by the admixture of unconscious psychic material.

It is widely stated that the alchemists failed in their endeavour to isolate the spirits. Before drawing such a conclusion we should be aware, more than we are, of the real state of affairs, hidden under the elaborate symbolism mentioned, concerning the influence of the elixir on the alchemist who consumed it. What did he understand by an "increased awareness" and did this have any relation to hallucinogenic effects? We still know at least one example of the spirits and their use. The alchemists were the first to distill wine and to obtain *aqua ardens*. This had wonderful properties. It could extract the quintessence of plants. It is on these lines that Dom Bernardo Vincelli, an alchemist, extracted and distilled a large number of herbs with alcohol. The heavenly influences which made them grow he combined into a "philosophical heaven" in the form of an agreeable liqueur, still consumed nowadays under the name of Benedictine[12]. There were less innocent spirits. In his *Alchemia* (1597)[22], Libavius, classifying, in the second part of his book, the products of a number of operations, extracts, distilled products, sublimates, etc., mentions potable metals, such as solutions of ferric acetate, mercuric nitrate, etc., and adds "but none will drink of this who is wise".

Besides the fact that the alchemists accumulated a mass of empirical observations and a number of technicalities (sublimation, distillation) which were the first processes of chemistry, the "macrobiotic" trend of alchemy developed in China, and later in the European context of the Renaissance, the first seeds of iatrochemistry, and paved the way for the development of the chemical concept of "pure substances", including those isolated from organisms.

All varieties of alchemy finally abandoned the laboratory to indulge exclusively in oratory. In the 18th century, for instance, Western alchemy finally turned into Hermetic Philosophy, which is outside our concern.

References p. 79

But another kind of metamorphosis influenced alchemy when certain alchemists of the 16th century, though persisting in seeing life and the Cosmos through a philosophical conceptual system, exchanged the stoic philosophy for Neoplatonism and Platonism.

8. Chinese iatrochemistry

Chinese alchemy, in its "macrobiotic" train of thought, developed the concepts of the "outer elixir" or longevity preparation made from inorganic substances or animal and plant constituents, and of the "inner elixir". The supporters of the latter maintained that important effects could be obtained only if an elixir of immortality was developed inside the body. This they endeavoured to produce, for instance, by the influence of gymnastic, sexual, or respiratory practices. In the iatrochemical movement which developed in China, from the 11th century to the 17th, the same concept was extended to the isolation of eventual components of the "inner elixir". In this way the Chinese iatrochemists prepared mixtures of crystalline samples of what were later called steroid sex hormones, using different varieties of these mixtures for the treatment of patients of different age or sex. Placenta, menstrual blood, testis and thyroid gland were also used as sources for the isolation of constituents of *materia medica*. Needham[23] rightly considers this as the greatest achievement of proto-biochemistry.

9. Paracelsus

As pointed out by Multhauf[24], the contributions of Fernel and of Telesio, referred to above, complicated the previous theories in two ways, by involvement of additional organs and a differentiation of the two biles into what we would designate as acid and alkaline. They nevertheless considered digestion as being due to heat, a concept rejected by Paracelsus and his followers, the medical chemists. Paracelsus introduced a rupture with ancient doctrines by assigning digestion not to "innate heat" but to "an internal chemist" (*Archeus*).

The medical chemists of the 16th century were the first to react against the Aristotelian theories. The systems of their predecessors were inspired by the contemplation of nature and embodied in rational discursive elaborations. They believed in the universality of the "elements", the relative proportions of which characterized each form of matter. The medical chemists embarked on "a search for our Creator through his created work by chemical investigations and analogies"[25], and they have rightly been considered to be the pioneers of a final phase of the scientific revolution[26]. Paracelsus (1493–1541)[27] represents the type of alchemist who turned towards chemical therapy and the preparation, not of gold, but of powerful drugs.

Paracelsus was one of those alchemists who preferred remedies taken from mineral origins to those endowed with symbolic or occult virtues derived from live sources. He was inclined to take a dynamic view of nature. A number of studies are at hand concerning the origins of his philosophy in Neoplatonism and Platonism. At different times, Paracelsus has expressed different views about the elements. He recognized four elements and four complexions. The elements are earth, water, air and heaven. One of the elements forms the quintessence of each object. Each of the elements possesses an *Archeus*. For instance the *Archeus* of water forms plants and trees from water. He recognizes an intermediary between the spirit and the matter of each of the three kingdoms: in minerals, the spirit, called Stannar or Truphat, produces crystal forms, whereas Leffos forms plants, and Everstrum, animals and man.

All things are generated from three principles: salt, sulphur and mercury, the last two borrowed from the alchemists. The three principles (*tria prima*) are for the form under which the four elements and the qualities (complexions) appear, sulphur representing the principle of combustibility, mercury the principle of liquidity, and salt the principle of fixity as resulting from the analysis by fire.

In the metabolic theory of Paracelsus which has analogies with the theory of Galen, digestion takes place in the stomach. The digested material is absorbed into the mesenteric veins, where nutrient is separated from waste (urine). The nutrient and urine are brought to the liver where the liver draws off its nutrients, and what is left is carried to the kidneys. (For a detailed analysis, see King[28].)

Farber remarks that Paracelsus has repeatedly attempted analysis by fire to distinguish between the forces of life and the substances of life. He introduced *Archeus* as a spiritual governor, but he also needed the concepts of a *spiritus vitae* and of *arcanum*.

"*Spiritus vitae* is a spirit which lies in all members of the body, as they are then named: and in all of them it is the same spirit, the one force in the one as in the other: And it is the highest kernel of life, from which all members live." (ref. 29)

References p. 79

Plate 9. Paracelsus.

Arcanum "is uncorporeal, undying, of eternal life." (Farber[29])

Quinta essentia "is a matter separate from all elements... and is a spirit alike to *Spiritus vitae*..." (Farber[29])

For Paracelsus, our constructions of forces or substances of life are valid only in the frame of the concept of the unity of life and of the universe. And yet the substance of life can be separated. The virtue of a metal or of a plant can be separated, the art of the chemist being necessary to separate the essence and leave the dead body behind.

"The substance of life can thus be prepared with almost no mass and with all the force-of-life." (Farber[29])

10. The Paracelsians

The chemical art of the followers of Paracelsus rested on his conviction that an active element of an *arcanum* could be identified with spirit. Along this line, the way was opened for the extraction of biochemical substances. This is the essential of the Paracelsian revolution: its instrument was the concept of the spirit.

"Because that which is visible, that is only the frame, and the element is a spirit, and lives the same in all things, as in the body: this is *prima materia* of the elements, invisible and unpalpable, and yet in all of them." (Farber[29])

The Paracelsians did not need to have deep philosophical theories to deal with in daily life. Spirits needed no explanation: they were, for the group of medical chemists, an obvious consequence of the unity of nature, as the existence of the spiritual soul of man is obvious for a Christian believer even today. The idea of the force-of-life having almost no mass (vital force, fifth essence, innate force) is more directly recognizable in poisons or in medicaments, acting upon life. The idea that this force-of-life can be separated into parts of matter, concentrated into the smallest quantity which can be used as an *arcanum* for a sick body, is the origin of the concept of a pure substance, and is not very different from the concept of "active substances" or "ergones" as sometimes used a few decades ago. Farber[30] remarks that historically the purification of alkaloids materially provided the first *arcana*. The force with almost no matter for its carrier of course corresponds, in a remote way, to our molecule of active substance. The concept was not only applied to poisons or drugs but also, for instance, to the "heat substance" (an ancestor of "phlogiston").

Ill-defined distillates and specifics with occult properties still constituted

References p. 79

Plate 10. Johan Baptista van Helmont.

the material bases of medical chemistry in the middle of the 16th century. During the second half of the same century, we find for the first time a *chemical* procedure in the production of drugs derived from metals used since the middle of the 14th century, such as vitriol (name of a variety of sulphates), mercury, antimony and gold. The process consisted generally of their solution in mineral acids, followed by a recovery in the form of inorganic salts. The introduction of these products into the pharmacopoeia was accomplished by the physician-chemists known as "Paracelsians".

Paracelsus used the "spirit of vitriol" as a medicine "*per se*", whereas his followers used it as a reactant for the preparation of mineral remedies. In this, the Paracelsians differed from Paracelsus in the fact that when he used a mineral acid in a reaction, he distilled the product and used the distillate, while they used the residue.

From John of Rupescissa to Croll we can follow the development of an extensive pharmacopoeia such as codified in Croll's *Basilica chymia*[31] in which a large number of chemical medicines that were in use during the 17th and 18th centuries are described[32]. The views of Paracelsians on metabolism are expressed by Croll as follows:

"The *Archeus* of man, or internal chemist, born within and implanted by God, when it takes to itself the food, at once separates the impure feculent tartar from the pure nutriment. Tartar is thus the excrement of food and drink... Out of the food, through the digestion of the *Archeus* in the stomach, that is, separation and generation of the separation, the primum ens comes to life from which comes the nutriment and food of the body. It is reduced into mercury (which occupies the place of nourishment), sulphur and salt, as is evident in the three principal waste substances." (translation by Multhauf[33])

11. Van Helmont

According to the scholastic (postgalenic) views of metabolism there were two pernicious biles, characterized by what we would designate as one with acid and one with alkaline properties accounted for by different organs, but evidently no reference was made to the still-unformulated concept of neutralization reactions. This was the merit of J. B. van Helmont (1579–1644)* who rejected Galenic as well as Paracelsian views of metabolism.

Partington[34] devoted a whole chapter (Chapter VI, pp. 209–243) to Van Helmont, and the reader is referred to this excellent biography, as well

* Van Helmont's works published up to the year of his death (1644) are listed by Partington[34]. His most important treatises are collected in *Ortus Medicinae*, published by his son in 1648 and of which several editions are known[35].

References p. 79

Plate 11. Franciscus Deleboë Sylvius.

as to Pagel's studies[36,37]. The main biochemical discovery of Van Helmont is that of acid as the digestive factor in the stomach, and Pagel claims that he also discovered hydrochloric acid as the agent of this digestion.

Van Helmont, without stating it, uses the law of the conservation of matter, and regularly uses the balance. He rejects both the four elements and the three principles, and considers air and water as the primary elements, finding the basis of this opinion in his famous willow tree experiment*. The ancient opinion was that air could be converted into water, but this Van Helmont did not believe. He observed, nevertheless, that water was converted into an air-like substance by evaporation, and this air-like substance easily returned to water. Van Helmont called it a vapour, observing that chemical reactions occasionally liberated air-like substances that could not be identified with air or with vapour, being more permanent than the latter. These air-like substances, subtler than vapour, denser than air, he called gases, the word gas being probably derived from the Greek term "chaos".

As well as Paracelsus, Van Helmont assumed the existence of "a chemist in the body" to direct its reactions. All reactions in life were, according to him, controlled by "ferments", *i.e.* that which converts other materials into its own nature. The "prima initia" of organisms are water and ferment of seminal origin (fermentum sive initium seminale), the ferment being an indwelling, individual formative energy, "hardly 1/8200 part of a body" which "disposes the material of water so that a seed is produced and life, and the mass develops into a stone, metal, plant or animal"[34]. By the action of the *Archeus*, as the effective cause, matter develops from within to certain forms:

"the seed is a substance in which the *Archeus* is already contained, a spiritual gas containing in it a ferment, the image of the thing, and moreover a dispositive knowledge of things to be done... one thing is not changed into another without a ferment and a seed".

* "I took an earthen vessel, in which I put 200 pounds of Earth that had been dried in a Furnace, which I moystened with Rain-water, and I implanted therein the Trunk or Stem of a Willow Tree, weighing five pounds; and at length, five years being finished, the Tree Sprung from thence did weigh 169 pounds, and about three ounces:But I moystened the Earthen Vessel with Rain-water, or distilled water (always when there was need) and it was large, and implanted into the Earth, and lest the dust that flew about should be comingled with the Earth, I covered the life or mouth of the Vessel, with an Iron plate covered with Tin (lamina ferrea, stanno obducta) and easily passable with many holes. I computed not the weigh of the leaves that fell off in the four Automnes. At length, I again dried the Earth of the Vessel, and there were found the same 200 pounds, wanting about two ounces. Therefore 164 pounds of wood, Barks and Roots, arose out of water onely". (From the translation of *Ortus Medicinae* by Chandler, 1662. Quoted from Partington[34].)

References p. 79

Plate 12. Daniel Sennert.

Van Helmont's seeds and ferments clearly have no relation to our enzymes, but it is apparent that they have a distant kinship with our concept of the gene, and of the mode of its partial transcriptions in different cell differentiations, an analogy which becomes even more striking if we remember that they correspond to the "rationes seminales" of St. Augustine, whose philosophy greatly influenced him. St. Augustine taught that God has deposited in matter a number of "rationes seminales" which, by successive germinations in the matrix of matter, "acceptis opportunitatibus", as opportunities presented, produced each a different species of corporeal being. The striking character of the analogy becomes even greater if we remember that Mendel, an Augustinian monk, was submitted to the same philosophical influences.

According to Van Helmont, food undergoes six fermentations in the organism. The stomach and spleen, working together, secrete an acid liquor producing the first digestion. The mass passes into the duodenum where it is neutralized by the gall of the gall bladder. In the mesentery and in the liver, the third digestion takes place. The nutritious chyle is absorbed by the veins and becomes blood, while the residue of the food progresses along the intestine, meets a stercoraceous ferment and becomes faeces.

The fourth digestion occurs in the heart where by the addition of the vital spirit, the blood turns yellow. The fifth digestion consists of the conversion of arterial blood into vital spirit, aided by a ferment present in the left ventricle. Some of it crosses the interventricular septum through minute pores. In each separate member, a separate ferment elaborates the nutritive principles from the blood. Van Helmont himself did not appear to confer any great importance on the mutual neutralization of acids and alkalis he had described, but this aspect became the starting point of the doctrine of a school of physicians, the iatrochemists, whose main figure was Franciscus Deleboë Sylvius.

12. Iatrochemistry

Sylvius accords the main role in digestion to saliva (ferment) acting in the stomach. Since he could attribute it neither to the "bitter" gall nor to the "insipid" saliva, he held the pancreatic juice to be the source of the stomach acid, and he decided that it was acid. The reasoning of Sylvius was based on analogies with experiments in inorganic chemistry interpreted in terms of "acid spirit".

References p. 79

Partington[34] facetiously calls him "a precursor of the modern pH cult". The dominant idea of pathology in the doctrine of Sylvius was that disease was caused by an "acidity" produced by an acid or an alkali which could be neutralized to cure the disease. Combustion and respiration were considered by Sylvius to be similar processes. He supposed that the heat of the blood came from the collision, in the heart, of the igneous particles in the acid spirit of the lymph and the alkaline salt of the bile (see Partington[34]).

Like the Paracelsians, many of the iatrochemists were seeking a new science based on chemistry, a new key of all nature based on chemical experimentation. It is true that they had a special interest in man, but there was a similar attempt on the part of many of them to apply chemical explanations to geocosmic and to macrocosmic phenomena.

REFERENCES

1 Ch. V. Langlois, *La Connaissance de la Nature et du Monde au Moyen Age d'après Quelques Écrits Français à l'Usage des Laics*, Paris, 1911.
2 K. E. Rotschuh, *Physiologie. Der Wandel ihrer Konzepte, Probleme und Methoden von 16. bis 19. Jahrhundert*, Freiburg and Munich, 1968.
3 Ch. Sherrington, *The Endeavour of Jean Fernel*, Cambridge, 1946.
4 B. Telesio, *De Natura Juxta Propria Principae*, Rome, 1565.
5 G. Bachelard, *La Formation de l'Esprit Scientifique*, Paris, 1947.
6 N. Sivin, *Chinese Alchemy. Preliminary Studies*, Harvard University Press, 1965.
7 A. Waley, *Bull. School Orient. Stud.*, 6 (1930) 11. [A citation from the 32nd section of the Pao Pu Tzu, 340 A.D.].
8 Ho Ping-Yü and J. Needham, *Ambix*, 7 (1959) 57.
9 H. J. Sheppard, *Ambix*, 6 (1957) 86.
10 G. Verbeke, *L'Évolution de la Doctrine du Pneuma des Stoïciens à St. Augustin*, Paris and Louvain, 1945.
11 A. J. Festugiere, *La Révélation d'Hermès Trismégiste*, 4 Vols., Paris, 1950–1954.
12 F. S. Taylor, *The Alchemists*, London, 1951.
13 J. Ruska, *Al-Rāzī's Buch Geheimnis der Geheimnisse*, Berlin, 1937.
14 G. Heym, *Ambix*, 1 (1938) 184.
15 F. L. Davis, *Isis*, 25 (1936) 327.
16 T. Geber, *The Works of Geber*, translated by R. Russell, London, 1678.
17 S. Trismosin, *Splendor Solis*, cited after Sheppard[18].
18 H. J. Sheppard, *Ambix*, 6 (1958) 140.
19 E. Waite (Ed.), *The Hermetic Museum Restored and Enlarged*, London, 1893, cited after Sheppard[18].
20 M. Maier, *Atalanta Fugiens; Hoc Est, Emblemata Nova de Secretis Naturae*, Oppenheim, 1618 [reprinted and edited by L. Wuthrich, Cassel-Basle, 1964].
21 C. J. Jung, *Psychology and Alchemy*, translated by R. F. C. Hull, 2nd ed., Princeton, 1968.
22 A. Libavius (A. Libau), *Alchemia*, Frankfurt, 1587 [reprinted and translated in German, Weinheim/Bergstr., 1964].
23 J. Needham, in J. Needham (Ed.), *The Chemistry of Life*, Cambridge, 1970.
24 R. Multhauf, *The Origins of Chemistry*, Cambridge, 1946.
25 A. G. Debus, *The Chemical Dream of the Renaissance*, Cambridge, 1968.
26 A. G. Debus, *The English Paracelsians*, London, 1965.
27 *Theophrastus Paracelsus Werke*, 5 vols., edited by W. E. Peuckert, Basel-Stuttgart, 1965–1968.
28 L. S. King, *The Growth of Medical Thought*, Chicago, 1963.
29 E. Farber, *Osiris*, 11 (1954) 422.
30 E. Farber, *Chymia*, 3 (1950) 63.
31 O. Croll, *Oswaldi Crollii Veterani Hassi Basilica Chymia*, Frankfurt, 1609.
32 R. Multhauf, *Bull. Hist. Med.*, 28 (1954) 101.
33 R. Multhauf, *Bull. Hist. Med.*, 29 (1955) 154.
34 J. R. Partington, *A History of Chemistry*, Vol. II, London, 1961.
35 J. B. van Helmont, *Ortus Medicinae*, Amsterdam, 1648.
36 W. Pagel, *Jo. Bapt. van Helmont, Einführung in die philosophische Medizin des Barock*, Berlin, 1930.
37 W. Pagel, *Bull. Hist. Med.*, Suppl. 2 (1944).

Chapter 3

Respiratory Theory of "Vital Heat" and Phlogistonic Proto-Biochemistry

1. Respiratory theory of "vital heat"

As we have recorded in previous chapters, motion was, in the first phases of proto-biochemistry, attributed to life-as-soul (*psyche*), the cause of life-as-action. We have noticed that Aristotle had conceived of an *innate heat* as indispensable to the functions of the body and its movement but that he did not link it with an interaction with air, considered by him as cooling the innate heat. For Galen, on the contrary, the *pneuma*, a mixture of fire and air, was essential to life and equivalent to life-as-soul. The importance of air as supporting the vital fire was emphasized by Fernel and other Galenists as well as by Paracelsus.

The "nitrous spirit", "nitrous substance", "nitro-aerial particles" of the air was a widespread concept at the time.

The discovery of gunpowder in the later Middle Ages had led to an analogy between chemical explosion and thunderstorms, and it was consequently assumed that the materials that went into gunpowder (sulphur and nitre)* accounted also for thunderstorms and for earthquakes. This view was confirmed by sensorial observation in the aspect of a sulphureous smell in volcanic eruptions as well as after thunderstorms (ozone). When it was observed, early in the 17th century, that the addition of salt lowered the temperature of an ice–water mixture, nitre (salpeter) was used for cooling purposes, and as it was also used as a febrifuge, its "frigorific property"

* Nitre (salpeter) is potassium nitrate, KNO_3. In gunpowder it is associated with carbon and sulphur. If heated, gunpowder explodes during the following reactions according to our present knowledge

$$4\,KNO_3 + 5\,C \rightarrow 2\,K_2CO_3 + 3\,CO_2 + 2\,N_2$$
$$2\,KNO_3 + 2\,S \rightarrow K_2SO_4 + SO_2 + N_2$$

References p. 96

Plate 13. Robert Boyle.

was clearly recognized. This property of nitre, thought to be present in air and playing a role in thunderstorms, was considered as explaining the formation of snow. Since nitre was also known as a fertilizer, an analogy was conceived between that property and the supposed increase of fertility of the soil under the influence of spring snow.

As indicated by Guerlac[1], when it was shown that salpeter (nitre) cured meat and prevented its decomposition, it was accepted that wintry air (full of nitre as shown by the production of snow) was particularly salubrious.

Debus[2] emphasized that the vital nitre is in fact a legacy of the gunpowder theory of thunderstorms and showed that the association of vital properties with salpeter had sprung from Paracelsian origins. Paracelsus and his followers considered life as a kind of combustion. As gunpowder produced flames when added to a fire, and as blowing up the fire with the air of bellows was by analogy considered to be due to aerial nitre, the combustion of life, ceasing when the inhalation of air in the lungs is stopped, was also related to aerial nitre.

It has been shown by Partington[3] that this view was widespread in the 17th century. Fludd, in 1633, formulated the view according to which the nitrous spirit of air is taken up in the blood[4].

That there is in the air a something (aerial nitre) which supports life as well as fire, was therefore the starting point for the experiments performed by the English physiological school of the mid-17th century[5-7] together with the deduction of Boyle, from his experiments, that a substance necessary to life is supplied by air. The history of the development of this conception in physiological studies is epitomized by Hall as follows:

"*(1)* Boyle's deduction from his experiments that air supplies a substance necessary to life; *(2)* Hooke's guess that this substance is present also in nitre and that it acts specifically as a solvent of the body's sulphur; *(3)* Lower's experimental demonstration that the brighter colour of arterial blood is due to something blood gets from air—something that he views as a "nitrous spirit"; *(4)* Willis' notion (not very different from Hooke's) that a "nitrous substance" from the air induces in living or burning bodies a volatilization of their sulphur; and *(5)* Mayow's rather different conception that volatilized sulphur dislodges nitro-aerial particles from the particles of air". (ref. 7)

Mayow developed the view into a coherent conceptual system.

"Mayow had demonstrated that "air is deprived of elastic force by the breathing of animals very much in the same way as by the burning of flame", and from this went on to assert that "we must believe that animals and fire draw particles of the same kind from air." (Mendelsohn[6])

These proto-biochemical statements should not be overvalorized. They are only the expression of indirect reasoning, based in analogies. For Mayow,

References p. 96

Plate 14. Thomas Willis.

the air enters the blood where it is deprived of its nitro-aerial particles. These, being mixed with the sulphureous (combustible) particles present produced a fermentation with effervescence and with a production of heat. The analogy refers to the fact that sulphureous minerals, when freshly dug and exposed to moist air, produce an effervescence and heat.

The views of Willis are of particular interest because he developed the theory according to which the degradation phase of the "dynamic permanence" is a combustion and provides the source of motion, as well as the idea that the subjugated contribution of the degradation phase provides the materials for the regeneration phase.

The analogy of life and flame had been formulated, as recalled in previous chapters, by many authors.

Even in our present perspective there are real analogies between an organism and a flame. To quote Dixon,

"Both consist of a region in which a complex series of successive chemical reactions is taking place: in both, heat is produced, largely by oxidation reactions; both have a shape or form which persists although the matter which composes them is constantly changing, so that there is a continuous flow of matter through them, and the atoms of which they are composed now are not those of which they will be composed some time later; both require for their maintenance in a steady state a constant expenditure of energy, and they disintegrate when this is cut off by starvation or by turning off the gas." (ref. 8)

However, as Dixon remarks, there are two fundamental differences in our modern perspective. First, in the flame the reactions take place because the temperature is very high, whereas in living matter, in which the temperature is low, the reactions occur because of the presence of specific enzymes ordering sequences of reactions. "In the flame you have chaos; in the cell, you have order"[8]. Second, in the flame the reactions are only breakdown reactions, whereas in the cell, as later developments will show, the reaction sequences include biosynthetic pathways.

Not only did Willis show that the source of heat was not friction, but he finally also excluded "fermentation"[9] which in 1659[10] was the key to his system, in line with Van Helmont's views. Willis considered that "the blood's sulphur particles erupt and unite with nitrous particles of the air" in a subjugated way, whereas for Mayow the volatilized sulphur dislodges nitro-aerial particles from air[7].

In spite of their inadequate nature, the studies of Boyle, Hooke, Lower and Mayow linked "animal heat", respiration and combustion in a single conceptual framework (see Mendelsohn[6] and Hall[7]). This led to the negation

References p. 96

Plate 15. Johannes Mayow.

of the "innate heat" and to the conclusion that atmospheric air was involved in the same manner in combustion and in respiration, and that breathing was the process responsible for providing animal heat. It must be pointed out that the authors just mentioned did not consider that anything may be given off in respiration. It was Black who noticed that "fixed air" (CO_2) was given off (see Guerlac)[11]. This discovery, nevertheless, was not generally accepted, and it had to cope with the new tendency of physiologists to find the substrate of their analogies not in chemistry, but in physics.

2. Iatromechanism

The chemical theory of life, which had been the starting point of the method of the iatrochemists and had its origins in the philosophies of Paracelsus and of Van Helmont, was repudiated and condemned during the last part of the 18th century under the influence of iatromechanism. The simulation of the animal to a machine originated from Aristotle's philosophy, in which organs of movement of animals had been compared with the "organa" of the war machines such as the arm of a catapult. For Aristotle, the principle of all movement is the soul, whereas Descartes, who took his analogies from the consideration of watches, clocks and automatons, explained the principle of movement as the relative dispositions of the parts of the machine, the model of the animal body or of the human body being itself found in automatons, which were very popular at the time. As emphasized by Steno in his *Discourse on the anatomy of the brain*[12], Descartes describes a machine which is able to perform the movements of man and does not describe man himself. Teleology is put aside by Descartes as the man he considers, and who is not the man of the anatomist, is built under the cover of God, but in reality the teleology is packed together at the point of origin.

In a very penetrating analysis of the concept of the relation of machines and organisms, Canguilhem[13] stresses a sentence of the *VIth Meditation* of Descartes, showing that he cannot explain the meaning of the construction of the animal-machine by God without doing so in terms of finality

"... Considérant la machine du corps humain comme ayant été formée de Dieu pour avoir en soi tous les mouvements qui ont coutume d'y être..."

In his *Description du corps humain*[14] Descartes states that the body (the animal-machine) only obeys the soul when it is mechanically prepared for it, the movements commanding each other as in the wheel-work of a watch.

References p. 96

The school of iatromechanists founded by Borelli was influenced by Descartes as well as by Galileo. Baglivi was one of its main adepts. He considers the body of man as being composed of pincers, hydraulic pipes, retorts, filters, sieves, bellows, springs, etc.

"Let us, he writes, put aside the great words of the chemists, such as "fusion", sublimation, precipitation, who try to explain the phenomena of nature and build a philosophy of their own; it is nonetheless indisputable that all these phenomena are only examples of the application of the laws of equilibrium, of those of the wedge, rope and spring and of the other elements of mechanics." (ref. 15, translation by author)

The success of iatromechanism rejected into oblivion the views of the iatrochemists and of their successors, and the chemical approach to life was abandoned until the advent of the phlogiston theory.

3. Friction theory of animal heat

The success of the iatromechanistic theories led the physiologists of the time to reject the respiratory theory of animal heat and to accept that heat had its cause in the motion of the blood through the vessels, an idea which can be traced back to Descartes and to Borelli. "Innate heat", according to the iatromechanistic school, arises in the blood, as a consequence of attrition and abrasion. Harvey himself linked the internal heat closely with blood. His theory of circulation consists of a cyclic process of the heating, evaporation and subsequent cooling and condensation of a fluid. For him the blood itself was the only *calidum innatum* and the heart received the "innate heat" from the blood. Hales also believed that animal heat is generated in the blood; it was also the opinion of Haller. The friction theory therefore became accepted by physiologists and by the majority of scientists. But the formulation of combustion in chemical terms was soon to revive with an emphasis on "sulphur" rather than on "nitre".

4. Phlogiston and life

"Sulphur", in the language of the Paracelsians, was the combustible part of the natural bodies, and it was accepted that combustible matters contained sulphur and belonged to the category of the "sulphureous bodies" such as oils, coals, metals and living organisms. But do these bodies containing the *principle of sulphur* contain the material *sulphur* which is solid and yellow? No, according to analysis. Should not the material sulphur be considered as a combination of the principle of sulphur (oil or phlogiston) with bodies conferring to it the other qualities, such as solid and yellow? Stahl (1660–1734), following this trail of reasoning, observed that different substances, in spite of being in many respects different, share the common property of combustibility. According to him[16], all combustible substances contain a common principle, "phlogiston", which escapes when the substance burns, and becomes visible as fire or flame (free phlogiston). Imprisoned in the structure of the combustible material, and forming a part of it, the phlogiston was liberated when the substance was disintegrated. For Stahl, combustion, which he related to fermentation, was the simplification of a complex matter formed by phlogiston and a material becoming incombustible when completely deprived of phlogiston. In both fermentation and combustion, the simplification was a result of pure movement. (For an extensive discussion of the general theory of phlogiston and of its origins, see Partington[17], Chapters XVII and XVIII.) Stahl believed that phlogiston was extracted from the air by plants, passed with the vegetable food into the animal body and finally returned to the atmosphere. Apart from this cycle, Stahl did not concern himself with the matter-of-life and forces-of-life. The chemists did not unanimously accept the phlogiston theory, and Boerhaave does not even mention it in his *Elementa chemiae*[18].

5. The concept of "heat capacity"

During the last decades of the 18th century, the important concept of *heat capacity* was introduced and its relation with *heat intensity* was defined. It was this advance in physical science that finally made possible the return and the experimental demonstration of a relation between combustion and respiration which had been discarded by the iatromechanists whose views had been almost generally accepted.

It had been known for a long time that an animal could not live where a fire could not burn. Black had not only arrived at the concept of latent heat, but had also devised methods for the comparison of heat capacities[6].

6. The pneumatic chemists

Before considering in greater detail the implications of the phlogiston theory of life, the work of the "pneumatic chemists" should be briefly recalled[6]. In 1757, Black showed that "fixed air" (carbonic acid) is given off in bubbles

References p. 96

Plate 16. Georg Ernst Stahl.

during fermentation and that it forms chalk with lime water. He also discovered that "fixed air" is continuously thrown off in the expired air during respiration. A student of Black, Cavendish (1731–1810) decomposed water, which consequently was no longer considered as an element. Priestley (1733–1804) discovered "dephlogisticated air" produced by the action of light on green plants. The following quotation, taken from his *Lectures on experimental philosophy*[19], shows with what deep insight Priestley conceived the biochemical importance of respiration and the role of dephlogisticated air (oxygen):

"It is the ingredient in the atmospheric air that enables it to support combustion and animal life. By means of it, most intense heat may be produced; and in the purest of it animals will live nearly five times as long as in an equal quantity of atmospheric air. In respiration, part of this air, passing the membranes of the lungs, unites with the blood and imparts to it its florid colour, while the remainder, uniting with phlogiston exhaled from venous blood, forms mixed air. It is dephlogisticated air combined with water that enables fishes to live in it." (Priestley[19])

Independently of Priestley, Scheele (1742–1786) had analyzed air and showed that it was mainly composed of nitrogen gas and "fire air" (corresponding to our oxygen).

Andrew Duncan, in a course of lectures delivered at the University of Edinburgh in the winter of 1774, extended the phlogiston theory to the biochemical field of the nature of animal heat. Duncan's theory is stated by Mendelsohn in the following sentences.

"Duncan's own hypothesis attributed animal heat to the evolution of phlogiston, or the principle of inflammability, contained in the blood. This process, which he believed to be brought about by the action of the blood vessels, rested on several points which he reported he had tried to demonstrate. First, the blood contains phlogiston which is evolved, extricated, or brought to a state of activity and motion by the action of the blood vessels to which it is subjected during the circulation. Further, "the evolution of phlogiston is a cause which, through nature, produces heat, whether that heat be apparently excited by mixture, fermentation, percussion, friction (or) inflammation." Lastly, the heat thus produced is equal to the highest degree of heat which animals in any case possess." (ref. 6)

Patricus Dugud Leslie, a student of Black, published in 1778 a book on the origin of animal heat[20]. Under his Christian name Patricus Dugud, he had written in 1775 a thesis on the subject[21]. The reader will find in Mendelsohn's book[6] an analysis of Leslie's theory, which can be summarized by saying that phlogiston is contained in the blood and released by respiration, the role of which is to carry phlogiston out of the body.

While Black thought that respiration was necessary for the generation of animal heat, Leslie considered respiration as a means of cooling the body

References p. 96

heat and of carrying off the excess of the phlogiston introduced by the nutriment. If the lungs acted by heating the blood, Leslie claimed that the organism would need less respiration in warm weather, which is in contradiction with common observation. In Leslie's work we find the concept of phlogiston considered as the nutritious matter to which the animal body owes its activity and vigour.

Crawford's *Experiments and observations*[22], which appeared a year after Leslie's book, are also based on the concept of phlogiston. The three propositions forming the basis of his theory of animal heat are summarized by Mendelsohn as follows:

"First, he proposed to demonstrate that atmospheric air contains more absolute heat than the blood passing from the heart to the lungs through the pulmonary artery (venous blood). Lastly, he believed he could prove that the capacity of a body for heat is reduced by the addition of phlogiston and increased by the separation of phlogiston." (ref. 6)

The first two propositions clearly show that Crawford correctly made a distinction between heat intensity (temperature) and amount of heat.

Concerning the changes taking place in air during respiration, Crawford leans on the teachings of Priestley and suggests that atmospheric air is converted into fixed air (precipitate in lime water) and phlogisticated air (extinguishing flames, noxious to animals). We remember that Priestley had shown that dephlogisticated air supports life five times longer than atmospheric air. Now Crawford measured the specific heats of atmospheric air, fixed air and dephlogisticated air (obtained from red lead), using bladders filled with the gas and immersed in water, and showed that the "absolute heat" of dephlogisticated air is 4.6 times greater than that of atmospheric air. This demonstrated Crawford's first proposition. The second (more absolute heat in the blood coming out of the lungs than in the blood entering them) he demonstrated by showing that arterial blood contains ten times more "absolute heat" than venous blood, in spite of the fact that it remains at the same temperature.

According to Crawford, atmospheric air which is inhaled into the lungs has a great amount of "absolute heat", whereas the blood returning from the periphery of the organism is rich in phlogiston. In the lungs, as the affinity of the air for the phlogiston is greater than the affinity of the blood for phlogiston, the latter leaves the lungs with the expired air. In combining with the phlogiston, the air gives up heat, which is taken up by the blood, the capacity of which for heat has been increased by the removal of phlogiston. In the process of circulation, the blood combining with the phlogiston gives out the heat received in the lungs. The extreme parts reached by circulation, having discharged phlogiston to the blood are in a state of increased heat capacity and take up the heat displaced by the invasion of the blood by phlogiston. The system of double attraction for heat and for phlogiston in metabolism is summarized by Partington and McKie[23] as follows:

Atmospheric air	+	venous blood	→	phlogisticated air	+	arterial blood
(air and heat)		(blood + phlogiston)		(air + phlogiston)		(blood + heat)

Comparing respiration and combustion, Crawford concludes, as did Black on the basis of the production of "fixed air", that they are similar, with the difference that in combustion the fire is separated from the air, whereas in respiration the fire is separated from blood. For Crawford, air contained fire or absolute heat, and it was the heat that was separated in the lungs to appear in arterial blood. This is what can be concluded from the 1779 edition of his book. In the edition of 1788 he modified his ideas on the line of Lavoisier's contribution, but he still retained the erroneous hypothesis he had formulated in 1779 that "dephlogisticated air" contains more heat than "fixed air".

On the other hand, results on the specific heats of gases by Crawford have been disproved by Delaroche and Bérard[24] who found the specific heat of oxygen to be lower than that of CO_2. As remarked by Partington[25], if he had made more correct determinations, Crawford would have concluded that the combustion of charcoal should produce cold. Crawford's trail of thought led to a blind alley and remained there.

7. Photosynthesis

A major achievement of the phlogiston era was the identification of the process of photosynthesis. Priestley had shown that plants thrive in air made "obnoxious" by animal respiration[26], and, in 1773, he was presented with the Copley medal of the Royal Society*. The President, Sir John Pringle, delivered the speech.

"... Neither candles will burn nor animals live" said Pringle in his speech, "beyond a certain time in a given quantity of air; yet the cause of either so speedy a death or extinction was unknown; nor was any method discovered for rendering that empoisoned air fit again for

* It is sometimes stated that Priestley was given the Copley medal for a method of making soda-water. This is nonsense. The medal was in recognition of the merits of the experiments recorded in his *Observations on different kinds of air*.

References p. 96

respiration. Some provision however there must be in nature for this purpose, as well as for that of supporting flame: without such the whole atmosphere would in time become unfit for animal life, and the race of men as well as beasts would die in pestilential distemper. Yet we have reason to believe, that in our day, the air is not less proper for breathing in, than it was above two thousand years ago; that is as far as we go back in Natural History. Now for this important end, the Doctor has suggested, to the Divine as well as to the Philosopher, two grand resources of Nature: the vegetable creation again is one, and the sea and other great bodies of water are the other... From these discoveries we are assured, that no vegetable grows in vain, but that from the oak of the forest to the grass of the field, every individual plant is serviceable to mankind; if not always distinguished by some private virtue, yet making a part of the whole which cleanses and purifies our atmosphere. In this the fragrant rose and deadly nightshade cooperate; nor is the herbage, nor the woods that flourish in the most remote and unpeopled regions unprofitable to us, nor we to them; considering how constantly the winds convey to them our vitiated air, for our relief, and for their nourishment." (ref. 27)

In 1777, Priestley returned to the subject and found that a "green matter" deposited on the walls of his water containers formed bubbles of "dephlogisticated air" when illuminated by the sun[28].

Ingen-Housz had been present at the meeting of the Royal Society when Pringle presented the Copley medal to Priestley. Pringle's speech aroused his interest in plant chemistry, and a few years later, in 1779, he published a book[29] in which the importance of light for "dephlogistication" was clearly demonstrated. He proved that plants do not merely improve the air by absorbing "mephitic air", as Priestley thought, but by producing "vital air" (oxygen). Rabinovitch[30] expresses the view of Ingen-Housz as follows:

$$\text{Air} + \text{light} \xrightarrow{\text{plant}} \text{something "phlogisticated" in the plant} + \text{dephlogisticated air}$$

Senebier[31], who discovered the role of CO_2 (fixed air) in photosynthesis, could have stated more precisely the theory of Ingen-Housz as follows:

$$\text{Fixed air} + \text{light} \xrightarrow{\text{plant}} \text{something "phlogisticated" in the plant} + \text{dephlogisticated air}$$

Ingen-Housz elaborated this theory in a new book[32] in which he furthermore adopted the views of Lavoisier and discussed the matter in terms of carbonic acid from the air, oxygen and light* in a theory that can be expressed as

$$\text{Carbonic acid} + \text{light} \xrightarrow{\text{plant}} \text{organic matter} + \text{oxygen}$$

8. Phlogistonic metabolic theory

To summarize the theory of metabolism adopted by the phlogistonists we

* The history of photosynthesis and of other aspects of biogenesis in plants will be retraced in Part IV of this History.

may state that, according to their views, the phlogiston or heat principle, is introduced into the organism with food, of which it constitutes a part, and is liberated by digestion. It is discharged into the blood in which it replaces the heat taken up from the blood by the parts of the body. Arriving at the level of the lungs, the venous blood abandons a part of its phlogiston to the pulmonary air with which it leaves the organism (in the form of phlogisticated air). In the lungs, the blood, while giving up phlogiston, takes up heat from the air and becomes arterial (blood + heat).

Plants take up phlogiston from phlogisticated air, exchanging it with heat and transforming phlogisticated air (air + phlogiston) into atmospheric air (air + heat).

The whole fabric of the phlogiston doctrine was to be overthrown by Lavoisier during what has been called the "chemical revolution".

References p. 96

REFERENCES

1 H. Guerlac, *Isis*, 45 (1945) 243.
2 G. A. Debus, *Isis*, 55 (1964) 43.
3 J. R. Partington, *Isis*, 47 (1956) 405.
4 G. A. Debus, *J. Hist. Med.*, 16 (1961) 374.
5 H. Guerlac, *Isis*, 51 (1960) 161.
6 E. Mendelsohn, *Heat and Life*, Cambridge, Mass., 1964.
7 T. S. Hall, *Ideas of Life and Matter*, Chicago and London, 1969.
8 M. Dixon, in J. Needham (Ed.), *The Chemistry of Life*, Cambridge, 1970.
9 Th. Willis, *Affectionum Quae Dicituntur Hystericae et Hypocondriacae...*, London, 1670 [cited after Hall[7]].
10 Th. Willis, *Diatribae Duae Medico-Philosophicae...*, 2nd ed., London, 1659 [cited after Hall[7]].
11 H. Guerlac, *Isis*, 48 (1957) 124.
12 N. Stenson, *Discours de Monsieur Stenon sur l'Anatomie du Cerveau*, Paris, 1669.
13 G. Canguilhem, *La Connaissance de la Vie*, Paris, 1952.
14 R. Descartes, *Description du Corps Humain*, Paris, 1664.
15 G. Baglivi, *Opera Omnia*, Leyden, 1704.
16 G. Stahl, *Theoria Medica Vera...*, Halle, 1708.
17 J. R. Partington, *A History of Chemistry*, Vol. II, London, 1961.
18 H. Boerhaave, *Elementa Chemiae, Quae Anniversario Labore Docuit, in Publicis, Privatisque Scholis Hermannus Boerhaave*, 2 Vols., Leyden, 1732.
19 J. Priestley, *Lectures on Experimental Philosophy*, London, 1794.
20 P. D. Leslie, *A Philosophical Enquiry into the Cause of Animal Heat, with Incidental Observations on Several Physiological and Chymical Questions, Connected with the Subject*, London, 1778.
21 P. Dugus, *De Caloris Animalium Causa*, Edinburgh, 1775.
22 A. Crawford, *Experiments and Observations on Animal Heat, and the Inflammation of Combustible Bodies, Being an Attempt to Resolve these Phenomena into a General Law of Nature*, London, 1779. [A second edition, in which Crawford took account of Lavoisier's work, appeared in 1788.]
23 J. R. Partington and D. McKie, *Ann. Sci.*, 3 (1938) 337.
24 J. P. L. A. Delaroche and J. E. Bérard, *Ann. Chim. (Paris)*, 35 (1813) 72.
25 J. R. Partington, *A History of Chemistry*, Vol. III, London, 1962.
26 J. Priestley, *Phil. Trans. Roy. Soc. (London)*, 62 (1772) 147.
27 D. McKie, *Ambix*, 9 (1961) 1.
28 J. Priestley, *Experiments and Observations Regarding the Various Branches of Natural Philosophy*, Vol. I, London, 1779.
29 J. Ingen-Housz, *Experiments Upon Vegetables, Discovering their Great Power of Purifying the Common Air in Sunshine and Injuring it in the Shade at Night*, London, 1779.
30 E. I. Rabinowitch, *Photosynthesis and Related Processes*, Vol. I, New York, 1945.
31 J. Senebier, *Mémoires Physico-Chimiques sur l'Influence de la Lumière Solaire pour Modifier les Êtres des Trois Règnes de la Nature et Surtout Ceux du Règne Végétal*, 3 Vols., Geneva, 1782.
32 J. Ingen-Housz, *An Essay on the Food of Plants and the Renovation of Soils*, London, 1796 [appendix to the 15th chapter of the General Report from the Board of Agriculture].

Chapter 4

"Dynamic Permanence" and "Assimilation" Prior to the Chemical Revolution

Prior to the second half of the 18th century, nutrients were considered necessary as lubricants for the motion of muscles and joints, and for replacing wear and tear. As we saw in Chapter 1, the process was considered by the ancient Greek philosophers as an addition or removal of small particles (concept of composition and decomposition, waste and repair). In the notion of direct assimilation, introduced by Empedocles, it is assumed that like attracts like, each part of the organism selecting from nutrients, without modification, the compounds of the size fitting the pores in the corresponding parts of the body. For Anaxagoras, assimilation was an aspect of his theory of "seeds", seeds of the substance of the body being present as such in the nutrients and selected by the parts.

In the writings of the Hippocratic Collection, food is considered as being carried to every part of the body where it is "assimilated", as a result of "coction" (consumption and repletion).

The atomists considered that the nutrients are broken up into their constituents, in the form of new atoms taken up by the organism and processed into the structure, a new atom taking the place of an old one.

Blood, for Plato, contained the substance needed to replenish the waste resulting from the assaults of the elements outside the body. As stated in the *Symposium*:

> "Even during the period for which a living being is said to live and retain his identity—as a man, for example, is called the same man from boyhood to old age—he does not in fact retain the same attributes, although he is called the same person: he is always in the process of becoming a new being by undergoing loss and restoration which affects his hair, his flesh, his bones, his blood and his whole body." (translation of Lamb[1])

The system of Plato is subject to simultaneous depletion (anachoresis) and repletion (plerosis) (from the blood).

References p. 104

According to Aristotle (in the *Metaphysics*)[2], Plato took the idea from his teacher Cratylus, who was Heraclitan. Aristotle himself expresses the same idea. He believes that the nutrients are concocted by vital heat into blood which brings to the body parts the modified individual constituents required by the nutritive soul.

For Galen, the nutrients produce lymph, which is transformed into humours supplying the body parts with the necessary material for the reparation of wear and tear. In the process the food undergoes an alteration which makes it similar to the substances of the organism. Blood formed in the liver undergoes different degrees of "coction" corresponding to the different organs. The blood contains the three other humours in order to nourish the varied members of the body. Galen differentiates the four natural humours in terms of their *complexio*: (blood, hot and moist; phlegm, cold and moist; choler, hot and dry; melancholy, cold and dry). He considers their origin in terms of fermentation, blood corresponding to wine, choler and melancholy corresponding to scum and dregs. Nutrition is defined by Galen in *De virtutibus naturalibus* (I, X–XI) as

"the assimilation of that which nourishes to that which is nourished. He speaks of assimilation as a mutual intrinsic change (*transmutare in se inuicem*), which is impossible unless the nutrient and that what is to be nourished already have "community and affinity in their qualities" (*nisi habeant communionem iam et cognationem in qualitatibus*). The process of nutrition is one of bringing the potential affinity present in the aliment to actual assimilation so that the body may be nourished." (Ogden[3])

For Galen and for the "humoralists", his successors, the blood was the theatre of all the events taking place in the organism, the organs being nourished by blood, through assimilation, but endowed only with the function of absorption and of secretion or excretion of the substances formed in the blood by virtue of its spontaneous activity.

Avicenna is clearer in his definition of the nutritive faculty in terms of composition (assimilation) and decomposition:

"that by which nutrient [matter] is converted into likeness of the nourished [member], so that by this restoration it [the member] may regain that which had been destroyed." (*Canon*, fol. 23 va, cited after Ogden[3].)

Guy de Chauliac, in his influential work *Chirurgia magna*, defines the natural humours as *materia nutricionis* by which he means that they contribute to nutrition by being specifically transformed into the body substance. For him the nutritional components of the blood constitute a special kind of natural humour[3].

Descartes[4] gives to the solid parts of the body a fibrillar structure:

"all the solid parts are composed entirely of differently extended, folded and sometimes interlaced fibrils (*petits filets*) each of which arises at some point on an arterial branch." (Hall[5])

He considers that the fibrils are continuously secreted by the arteries. At each beat of the pulse, arterial pores are opened. Through them blood particles are ejected and strike the roots of the fibrils. Consequently, when a particle is detached at the end of a fibril, another is attached at its root.

Boyle also proposes a mechanical model for assimilation and proposes that, from the blood,

"the corpuscles of the juice insinuate themselves at those pores they find commensurate to their bigness and shape; and those that are the most congruous, being assimilated, add to the substance of the part, wherein they settle, and so make amends for the consumption of those that were lost by that part before." (ref. 6)

We may also mention here the vital motions postulated by Willis[7] to which is attributed the encounter of arterial juice and nerve juice, as well as assimilation.

The concept of assimilation, although generally limited to living organisms, has sometimes been valorized to such a degree as to become inserted into a cosmic perspective. Hunault[8], for example, puts on the same level the "indigestion" taking place in the earth, in the kitchen and in the stomach.

Bourguet[9] adheres to the concept of the "third coction" (the third of the successive digestions in the stomach, the liver and the parts of the organism). He makes a distinction between the parts growing by an addition of matters through layers added to the exterior surface, as in crystals (a method he recognizes in non-living parts such as hair and nails), and living parts growing by "addition of new molecules in the whole of the inside at the same time" ("intussusception" in the language of the schools). He suggests the existence in the bones, nerves, membranes, muscles, etc., of

"organs so admirably built, that they are able to transform the humours they receive from the general mass of the blood and to appropriate them to their own substance by a mechanism which is the same in all, and peculiar to each". (ref. 9, translation by author)

Here again we find the "third coction" in the form of a faculty of assimilation. From the Greek atomists, Bourguet borrows the "mould" or "matrix" in which the matter-of-life is cast from chyle*.

The "organic mechanism" of Bourguet was opposed by Colonna[11], who, in spite of being an atomist, criticized mechanism and considered that the "moulds" existed only in the minds of their authors. On the other hand,

* On Bourguet, see Roger[10].

Plate 17. Herman Boerhaave.

it appears absurd to Colonna that a plant may select particles on the basis of their adaptation to the form of the openings in the roots, as particles small enough to go through will do so whatever the form may be. This criticism was also formulated later by Bordeu.

Boerhaave considers the decomposition aspect of the dynamic permanence ("decrement") as mechanical. His "increment", or recrement, he considers as being due to the faculty of converting food into a matter similar to that of the organism, and

"fit for augmenting or restoring such parts of the body as are decayed or consumed." (Boerhaave[12])

As stated by Hall[5], Haller follows a similar concept.

"He saw fluids as "squandered" by way of the breath, by tears and other secretions, and by sweat, urine and feces. The erosion of solids was mechanical rather than chemical, he supposed, being caused by pulse beats (small but numerous), by the friction of the body fluids, by abrasion at the edge of loose membranes, and by the wear-and-tear of muscular movements. Displacement (of solids) is a crumbling that occurs when the parts come unglued." (ref. 5)
"A gelatinous juice is conveyed from the aliments through the arteries to all parts of the body and exudes into the cellular texture everywhere." (translation edited by Cullen[13])

For Needham[14], life consists of a certain degree of exaltation of matter. The most "exalted" substances assimilate the less "exalted", as a plant or an animal assimilates nutrients, fire "assimilates" wood, spark assimilates powder, acid assimilates metals. The overvalorization of the concept is here as marked as in Hunault's[8] views; in fact it is a vitalism in which the whole of nature is conceived on the model of vital phenomena.

The concept of the "mould" is also part of the system of assimilation formulated by Buffon[15]. It is not possible to deal here, whatever their interest may be, with the implications of the concept in ontogenesis or in phylogenesis. For Buffon, the "interior mould" is one of the components of the eternal order of nature, such as gravitation for instance. But the "interior moulds" are not general effects, they are particular effects. For Buffon, at the origin of life, "organic molecules" were formed as they entered into combination to form animals and plants, by which we mean not only those we know today, but all the possible combinations. Some have persisted, others have disappeared. Evolution, for Buffon, is a selection of a number of forms which have persisted when other forms have disappeared. For Buffon, the same kind of "nutrition" exists in all creatures, as well as of "development" and "reproduction". The unity of life is stressed. Buffon, by the notion of "interior mould", gives to matter a certain spatial disposition. The "mould" itself absorbs the assimilated matter, but the

References p. 104

Plate 18. Albrecht von Haller.

intimate incorporation is only accomplished by the action of "penetrating forces" inherent to the interior of matter. Each part of an organized body, each "interior mould", only accepts the "organic molecules" which belong to its nature. The "living organic molecules" of Buffon, existing in the whole universe, formed by matter-of-life, and existing as well as the non-living molecules, were apparently formed at the time of the origin of life, although it is difficult to determine the origins of the concept.

The idea of a primary living matter, existing as such in the universe, and used by the organisms to form their own substance, has been accepted by several authors. Bonnet[16] contradicts Buffon on "organic molecules", to which he refuses to recognize the power of forming the organism by their "penetrating forces". For him, the nutrients produce nothing by themselves, and they simply become, by "assimilation", constituent parts of the organic whole.

Bonnet considers assimilation as an incorporation by intussusception, a suction of juices into the interstices of the fibres, which grow in all directions. He adheres to the concept, often noted in the previous pages, that the ultimate use of nutrients is preceded by assimilative changes forming substances analogous to the nature of the body.

In the *Rêve de d'Alembert*[17] of Diderot, first published posthumously in 1830, we hear Melle de Lespinasse and Bordeu at the bedside of the sleeping d'Alembert, discussing the questions he raises in his dreams. When he awakes and asks how, in spite of the continuous replacement of matter, he remains himself (a new aspect of the concept of dynamic permanence), Bordeu answers by comparing the organism to a monastery in which new monks arrive and others leave, in spite of which the monastery keeps its integrity, a concept that wel shall later find (Chapter 10) in more modern formulations.

In all the theories referred to in this chapter, food is considered as bringing directly or after slight modifications, to the parts of the organism, the matter necessary to repair wear and tear. Moreover, no relation was suspected between food on the one hand, and force or heat on the other hand.

References p. 104

REFERENCES

1 Plato, *Lysis, Symposium, Gorgias*, W. R. M. Lamb (Ed.), Loeb's Classical Library, London and Cambridge, Mass.
2 Aristotle, *Metaphysics*, 2 Vols., H. Tredennick (Ed.), Loeb's Classical Library, London and Cambridge, Mass.
3 M. S. Ogden, *J. Hist. Med.*, 24 (1969) 272.
4 R. Descartes, *Oeuvres*, Ch. Adam and P. Tannery (Eds.), Paris, 1897.
5 T. S. Hall, *Ideas of Life and Matter*, 2 Vols., Chicago and London, 1969.
6 R. Boyle, *An Essay on the Porousness of Bodies*, 1684.
7 Th. Willis, *Cerebri Anatome, cui Accessit Nervorum Descriptio et Usus*, London, 1664.
8 Hunault, *Discours Physique sur les Fièvres qui ont Régné les Années Dernières*, Paris, 1696.
9 L. Bourguet, *Lettres Philosophiques sur la Formation des Sels et des Crystaux et la Génération et le Mécanisme Organique des Plantes et des Animaux*, Amsterdam, 1729.
10 J. Roger, *Les Sciences de la Vie dans la Pensée du 18e Siècle*, Paris, 1963.
11 F. M. P. Colonna, *Histoire Naturelle de l'Univers*, 4 Vols., Paris, 1734.
12 H. Boerhaave, *Elementa Chemiae, Quae Anniversario Labore Docuit, in Publicis, Privatis que Scholas, Hermannes Boerhaave*, 2 Vols., Leyden, 1732.
13 A. von Haller, *First Lines of Physiology*, Reprint (with an introduction by L. S. King), New York and London, 1966.
14 J. T. Needham, *An Account of Some Microscopical Discoveries*, London, 1745.
15 G. L. Leclerc de Buffon, *Oeuvres Complètes, Augmentées par M. F. Cuvier*, Paris, 1829–1832.
16 Ch. Bonnet, *Contemplation de la Nature*, 2 Vols., Amsterdam, 1770.
17 D. Diderot, *Oeuvres Complètes*, Vol. 2, Paris, 1875.

Part II

From Proto-Biochemistry to Biochemistry

Chapter 5

Metabolic Theories of Lavoisier and his Followers

1. Lavoisier on combustion and respiration

As shown in Chapter 3, the fact that a substance provided by air was necessary for life and for the production of animal heat had been foreseen by Galen and by Paracelsus, and this had been accepted by Boyle as well as by the group of English physiologists of the mid-17th century, but had been discarded by the iatromechanists.

Lavoisier's work and his numerous contributions to the progress of chemistry have been the subject of extensive studies. Although most of them are outside our interest, mention must be made of his contributions when they led him to the study of living organisms and when we can recognize the trail of the research he followed in this direction. Lavoisier was a wealthy financier actively interested in the progress of industry and agriculture. He had been active in developing systems for lighting the streets at night, and this interest led him to study the chemical phenomenon of combustion, which led to respiration.

In 1772 he started a long series of experiments on calcination and combustion. (For the origin and background of these experiments, see Guerlac[1].) He pursued these studies with a knowledge of the work done on the chemistry of gases by Hales, Black, Cavendish and Priestley in England. At the end of 1772, Lavoisier reported that when phosphorus burns, it combines with air and produces phosphoric acid (acid spirit of phosphorus) weighing more than the phosphorus at the start. The same result, increase of weight, was also observed when sulphur burns: it combines with oxygen, and the sulphuric acid (vitriolic acid) produced weighs more than the original sulphur. At this early stage he also realized that when metals are calcinated, they also combine with air and increase in weight. These brilliant discoveries were in complete contradiction with the teachings of the phlogiston theory,

References p. 126

according to which combustion or calcination were accompanied by the loss of an immaterial constituent and not with an increase of matter.

In the year 1777 Lavoisier, in two papers read at the Royal Academy of Sciences[2,★], gave precise details of the elementary ideas concerning his conception of animal heat. He came to the tentative conclusion that respiration is of the nature of a combustion and capable of generating animal heat, as had been part of the teachings of Black. From his experiments Lavoisier concluded that one-sixth of the air is respirable ("pure air", corresponding to the "dephlogisticated air" of Priestley) and able to support combustion and the calcination of metals. Experimenting with guinea-pigs and with birds, Lavoisier concluded that his "pure air" is converted into "acide crayeux aériforme" (fixed air, CO_2). In Lavoisier's papers we find a clear explanation of some observations of Priestley. Lavoisier confirms that air modified in the course of the transformation of mercury into its oxide is inapt to serve in respiration, as is air modified by the respiration of a bird (two forms of "mephitic air"). But Lavoisier shows that this mephitic air contains "fixed air" (CO_2). Removing from this mephitic air the "fixed air", which does not exist in the other form of mephitic air, Lavoisier obtains identical gases which he calls "moffette" (he later called it azote, and Chaptal, in 1790, called it nitrogen, corresponding to the mephitic air of Rutherford). To complete the experiment, Lavoisier restitutes oxygen to each of the two "moffettes" and renders to the air its capacity to serve in respiration. Two explanations could be given for the production of "fixed air" (acide crayeux aériforme, CO_2) from "eminently pure air" (which Lavoisier called oxygen in 1779). Either oxygen is actually converted into CO_2 by a local combustion, or it combines with the blood, in an exchange with CO_2. For Lavoisier, oxygen was not a free element but was either combined with the "heat substance" as a gas, or with a metal, in calcination, after the release of the heat substance (caloric).

Lavoisier's work, in a later development, brought the concept of the relation of respiration and animal heat outside the field of analogies with the chemical phenomena of combustion, to place it in the field of the consideration of the processes actually at work inside the animal body. In a paper he published with Laplace[3] in 1783 the theory of animal heat is based on the first direct measurements of the heat dissipated by a living organism. As stated by Berthelot[4], Lavoisier and Laplace were bold enough "to compare a living body to a chemical compound".

★ The volume of the *Mémoires* for 1777 was distributed only in 1780.

Lavoisier and Laplace[3], after an introduction showing that they make no distinction between the biological and the physico-chemical worlds, state again the hypotheses formulated by Lavoisier in his previous work, that heat is evolved in any decomposition of "pure air" and that "fixed air" (CO_2) is produced in respiration as well as in combustion, the heat resulting from the decomposition of "pure air" being the source of animal heat. The authors clearly define the difference between their conception and the ideas of Crawford. While Crawford sees heat as being free "in pure air", only disengaging itself when "pure air" loses some of its capacity for heat, Lavoisier and Laplace believe that the heat comes from the "gas oxygen" which they consider as a chemical combination between oxygen and heat (caloric). This is another way of stating that the heat results from an exothermic reaction of oxidation. The authors, using a balance and an ice calorimeter, measured the effects of the respiration of an animal and compared it with the combustion of a candle. They first determined, with a balance, the quantity of CO_2 produced by a guinea-pig freely respiring in air for a known time. Then they placed the animal in their ice calorimeter and measured the heat liberated during a period of ten hours. The authors claimed that their results showed that in combustion as well as in respiration, the production of equal amounts of CO_2 was accompanied by a production of equal amounts of heat. Respiration, they concluded, is a slow combustion.

Lavoisier and Laplace clearly enunciate their theory of respiration as follows:

"Respiration is thus a combustion, very slow it is true, but perfectly similar to that of carbon; it occurs in the interior of the lungs, without disengagement of visible light since the matter of fire which becomes free is at once absorbed by the humidity of these organs. The heat developed in this combustion communicates itself to the blood which traverses the lungs and from there it is spread over all the animal system". (ref. 3, translation by Partington[5])

Seguin and Lavoisier situate the combustion in the cavities of the tubes of the lung.

"One must know in the first place," they write "that there transudes into the bronchi a humour which is secreted from the blood and which is principally composed of carbon and hydrogen." (ref. 8, translation by Foster[6])

To quote Holmes

"Lavoisier viewed respiration as a simple, direct combustion of carbon and hydrogen furnished by the blood, analogous to the burning of oil in a lamp." (Holmes[7])

Seguin and Lavoisier[8] logically suggested a proportionality between respiration and digestion as the carbon and hydrogen continuously afforded for

References p. 126

Plate 19. Antoine Laurent Lavoisier.

combustion in the cavity of the lungs has to be replenished from the food. It was the first statement in history, of a relation between food on the one hand and animal heat on the other. But this was an idea without any experimental basis:

"If the animals did not compensate by the nutrients what they lose by respiration, the lamp would soon lack oil and the animal would die as a lamp burns out when it lacks fuel." (ref. 6, translation by author)

The theory according to which respiration occurs in the interior of the lung remained the current view during half a century, in spite of a number of successive attacks. We shall return to this question when dealing with cellular oxidations.

2. The chemical revolution

As far as Lavoisier's work is concerned outside the field of metabolism, we may refer the reader to the chapter on Lavoisier in Partington's *History of Chemistry*[5] (Vol. III, Chapter IX).

The "chemical revolution" is primarily the work of Lavoisier. His entirely new conception of combustion resides in the combination of oxygen with other bodies whereas his predecessors had considered it as a degradation process. The combustion of hydrogen results from its union with oxygen and produces water. In his system, respiration is a slow combustion using oxygen and producing carbon dioxide. The nature of water and of carbon dioxide being known, this leads to the knowledge of the composition of organic matter constituting the organisms. As their combustion, as well as their distillation, produces water and carbon dioxide the organisms appear to be composed of carbon, hydrogen and oxygen; and soon afterwards the presence of nitrogen was demonstrated in their constitution. This embodies a deep rupture with the doctrine of the four elements as well as with the theory of phlogiston. In one of his early works, Lavoisier contradicted two generally accepted theories: (*1*) that water could be converted into earth by repeated distillation; and (*2*) that it was possible to convert water into a gas so "elastic" as to pass through the pores of a vessel.

This disposed of the idea of the interconversion of the four elements.

After his work on combustion and respiration, Lavoïsier was in a position to lead a decisive attack on the phlogiston theory. In this he was supported by the work performed in England on the composition of water. He followed up that line and showed that water was a compound of hydrogen and oxygen.

In the *Méthode de Nomenclature Chimique* (1787)[9] Lavoisier, with Guyton de Morveau, Berthollet and Fourcroy, proposed a method for naming chemical compounds. It is based on his concept of oxygen as the universal acidifying principle, which combines with carbon to form carbonic acid; with sulphur to form sulphuric acid; with nitrogen to form nitric acid; and with metals to form oxides.

In order to ensure a final victory over the theory of phlogiston, Lavoisier published in 1789 his *Traité Élémentaire de Chimie*[10] in which was presented the first table of chemical elements,

"a table of simple substances belonging to all kingdoms of nature, which may be considered the elements of bodies." (ref. 10, translation by author)

In the *Traité Élémentaire de Chimie*, Lavoisier states clearly and explicitly the principle of the conservation of matter.

In the course of his researches, Lavoisier conceived the first methods of organic analysis with the help of which he studied the composition of several organic compounds.

3. Proximate principles

No scientific biochemistry could have developed of course before chemistry, under Lavoisier's influence, became a distinct and coherent conceptualist structure. In the period preceding this retarded emergence, three main intellectual processes did go through changes, defined by Farber[12] as emergence from deification, emergence from humanization, and emergence from universality. The emergence from deification, in which the substances acquired independence from the influences of gods and planets, coincides with the abandonment of alchemistic ideas in the 18th century. The Greek philosophers started the emergence from humanization in the 5th century B.C. The speculations about motion and about particles of matter, which started in the 13th century, prepared its end, as well as the mathematical abstractions introduced by Newton and by Descartes in the 17th century. In the course of the 18th century, the universality of principles, present in all forces of matter, originally invoked to account for the continuous flow of changes, was replaced by the recognition of distinct substances. Of the concepts of earth and fire and water as elements of all substances, none stood the test of the 18th century. Earth had ceased to be regarded as an element when air was shown to be a mixture of chemical entities.

Nevertheless, at the end of the 18th century, and even in the beginning of the 19th century, matter-of-life is often still directly studied in bulk. Lavoisier, in 1788 still refers to distillates of meat, wax, etc.; and Fourcroy, as well as Berzelius, still directly studies complex materials from organisms. This is still in line with such a 17th century book as the *Traité de Chimie* (1660) of Nicolas le Fèvre[13] who, in the first part of the first volume, deals with honey, wax, manna, mummy, blood, urine, and with their distillation. Nicolas Lemery, in the first edition (1675) of his *Cours de Chymie*[14], also still describes vipers and their distillation, as well as the distillation of honey and wax.

It is only after the formulation of the concept of the molecule that the chemists began to draw up the inventory of the chemical species, the association of which forms the substance of organisms; and it may be said that no scientific biochemistry could have developed before the concept of the molecule was defined. The concept of the chemical atom was introduced in 1803 by John Dalton who, in 1808, published the first volume of his *New System of Chemical Philosophy*[15]. In 1811, Avogadro made the distinction between the atom and what he called the compound atom, composed of two or more elementary atoms and to which he proposed giving the name molecule. Although, at the beginning of the 19th century, the chemistry of the metals and of the elements such as sulphur and phosphorus was much better known than the chemistry of carbon compounds, many definite compounds were recognized as being part of organisms. A prime source of bibliography concerning the studies mentioned below is afforded by Partington[5] (pp. 78 ff.). Rouelle investigated lactose (1773), starch (1770) and chlorophyll (1773). Scheele discovered tartaric (1770), mucic (1780), lactic (1780), uric (1780), prussic (1782–83), oxalic (1776, 1784–85), citric (1784), malic (1785), gallic and pyrogallic (1786) acids, murexide (1780), glycerol (1783–84), casein (1780). Fourcroy and Vauquelin obtained benzoic acid (1799) through the action of hydrochloric acid on horses' urine, in fact from hippuric acid, and of this Dessaignes later identified the other component as glycine (1846). In 1802 they investigated formic acid which they incorrectly considered to be a mixture of acetic and malic acids (the individuality of formic acid was established by Suersen

References p. 126

in 1805). Berzelius discovered lactic acid in meat juice in 1807, and xanthophyll in leaves in 1837. From ox bile he obtained what he called cholic acid (1841); he prepared pure casein from milk in 1814 and fibrin from blood in 1813. In bile, Tiedemann and Gmelin found cholesterol (1827). Before 1820, Pelletier and Caventou isolated a number of alkaloids: strychnine, brucine, quinine, cinchonine, veratrine. Robiquet isolated asparagine (1805), narcotine (1817) and codeine (1832). Braconnot found ellagic acid in galls (1818), pectic acid in fruits (1825) and equisetic acid in *Equisetum fluviatile* (1828) (later shown by Regnault, in 1836, to be identical with maleic acid discovered by Lassaigne in 1819).

This trend of research was rationalized through the introduction of two important concepts. The first was the concept of combination contrasted with mixture. The concept of combination was expressed as Proust's law of definite proportions, but it was only accepted after about eight years of discussion with Berthollet[16].

After introducing the notion of chemical species, Chevreul[17,18] formulated the methods for identifying what he called "principes immédiats" (an expression first used by Fourcroy[19] and which can be translated into "proximate principles") by the "analyse immédiate" (distinct from the "analyse élémentaire") and consisting of the isolation, without modification, of the chemical species as they exist in the living organism. Chevreul is rightly considered as the founder of biochemical analysis[18]. He had always been interested in the relation of crystal forms, or of colours, with the arrangement of elements in their combinations, and he had a special inclination for esthetic aspects such as tapestry dyeing or the impact of pigments in a certain order assembled on a canvas. He discovered the "simultaneous contrast" of colours which was so brilliantly taken advantage of by a host of modern artists, such as Delaunay or Herbin. Living as he did in the "Jardin du Roi", he was highly impressed by the use the naturalists had made of the species concept.

Chevreul's interest in fats was initiated in the frame of technological aspects. It had been known for a long time that it is possible to make soap by boiling fats with potash. In 1809[20], when examining a soap that had been successfully used for the fulling of cloth, he observed the presence of crystals in the sample he had to analyze. These crystals he recognized as formed by what he called potassium margarate* a name he coined to recall the pearly luster of these crystals. As noted above, Chevreul[15] thoughts

* They were later recognized as a mixture of stearate and palmitate.

were oriented by the purpose of introducing into chemistry the concept of the species, which had been so successful in biology, and which had already inspired Lavoisier[10]. He saw that mineral acids decomposed the crystalline soap into salt and an organic acid which he could crystallize out of the solution. When he had recognized that the saponified fatty substances were real acids, he was able to isolate them from the biological material in the crystalline salt form, by combining them with alkalis. He devised the criterion of purity provided by repeated crystallizations: if they were not followed by a change in melting point, the purity was established. In the lyes of most fats a sweet principle was recognized which Chevreul called "glycerine" (our glycerol). From the various fats, he obtained acids showing differences in melting point[17]. He concluded that saponification, the process of making soap, is a fixation of water, glycerol being formed by the combination of some elements of the fats with elements of water. A fat is comparable to a salt, and in saponification an inorganic salt-forming base takes the place of glycerol. The proximate principles of the fats (at least of those we call glycerides) are therefore glycerol and fatty acids:

"The old philosophical principle of oiliness or fattiness as a substance is thus replaced by a principle of composition or chemical structure." (Farber[12])

In the course of this work, Chevreul devised the method of chemical analysis applied to the constituents of the organisms[18] and recognized a "proximate analysis" (analyse immédiate) whose object is to separate the substances (organic species) composing the material: this differentiates it from "elementary analysis" which determines the nature and proportions of the elements composing the proximate principle.

In a treatise published in 1826, Hünefeld[21] considers, according to Chevreul's views, the results of elementary analysis and of proximate analyses. In the first volume, he enumerates a number of proximate principles (albumin, mucin, fats, lactic acid, etc.). In the second volume, Hünefeld expresses his views on metabolism, and for the first time he considers the phenomenon of the biosynthesis of proximate principles. The nitrogenous constituents of tears, of saliva, of sperm, of the crystalline lens, he considered as being different chemical species, but all derived from albumin. He believed that between the act of chymification and the acts of secretion and excretion, the substances undergo a succession of changes. Albumin is formed during the acts of "chymification" and "chylification", whereas fibrin is formed during "sanguification" and "respiration". From albumin,

References p. 126

Plate 20. Michel Eugène Chevreul.

the nitrogenous substances of the tears, of the crystalline lens, etc. are formed in the respective organs where they are found, or secreted. Urea and lactic acid are derivatives of the residues of the formation of "animal parts", and from urea, uric acid is derived. The very useful quality of the concept of proximate principles is to be contrasted with the theories of those authors who claimed that the nitrogenous components of living matter are undefined compounds, as in Raspail's *Système de Chimie Organique*[22] in which utter confusion reigns. In 1835 Wöhler wrote to Berzelius:

"organic chemistry appears to me like a primeval forest of the tropics, full of the most remarkable things." (ref. 23)

The efforts of many chemists had to pave the way for the clear recognition of the nature of the "chemical species" forming the substance of organisms. At the end of the 18th century, many substances obtained from organisms were in common use. They were known with respect to origin and use: but the chemists, with the help of the methods for elementary analysis originated by Lavoisier, began collecting data on the empirical formulas of organic natural substances, in a critical way. For instance, Dumas found the formula $C_{40}H_{52}O_5$ for oil of cloves, but Liebig and Ettling showed it to be a mixture of terpene $C_{10}H_{16}$ and an acidic substance which they called acid of cloves (Nelkensäure), $C_{24}H_{20}O_5$. Dumas called it eugenic acid and determined its elementary formula as $C_{20}H_{24}O_5$ (correct formula: $C_{10}H_{12}O_2$). Tiemann[24] showed that eugenol is a phenol $C_6H_3(OH)$-$(OCH_3)C_3H_5$.

4. The concept of animalization

While the knowledge of the nature, properties and elementary composition of natural substances (empirical formula) was increasing, chemists developed the concept according to which chemical changes were due to the addition or subtraction of one of the elements, or to recombinations of those elements. (For a detailed review of these metabolic theories based on elementary analysis, see Holmes[25].) As "organic chemistry" was at that time considered to be synonymous with our biochemistry, the chemists also applied the concepts that have just been defined to the changes taking place in the organisms themselves and particularly to the concept of dynamic permanence. Their theory had a kinship with the views of the Greek atomists on composition and decomposition which they also located at the level of the substances.

References p. 126

Plate 21. Antoine François Fourcroy.

The first definite attempt at explaining "composition and decomposition" chemically was made by Fourcroy[26] who, in 1789, concluded that the conversion of nutrients to the body substance of an animal consisted of an addition of nitrogen to the vegetable substance. This was an expression of the faith of the chemists of the time in the value of characterization of the organic substances through their empirical formulas. Hallé published in 1791[27] a scheme according to which nutrients have in common a base composed of carbon and hydrogen to which carbon and hydrogen were added in different amounts. Vegetable nutrients, of similar composition, were transformed into animal substances (animalized) in successive steps, each consisting of an addition of nitrogen and a removal of carbon (combined with oxygen to form CO_2) and taking place respectively in the intestine, in the lungs and at the level of the skin. The theory of *animalization* by increase of nitrogen content, after Fourcroy and Hallé had proposed it, became widely accepted, and it is presented by Huenefeld in his book[21]. Richerand in 1807[28] specified that gelatin was successively transformed by animalization into albumin and finally to "fibrin". Many authors had expressed the view that the fibrin of blood was directly transformed, by simple solidification, into muscle "fibrin".

As they were soon discouraged by the difficulty of defining all the compounds concerned, the chemists turned to the analytical study of bulk foods, body fluids and solids, and excrements. Gmelin[29] formulated a theory of metabolism in plants and animals, based on elementary analysis and according to which plants increase the complexity of organic compounds by reduction, whereas animals degrade them by oxidation. This became generally accepted, with the result that plants were considered by the school we may designate as "the analysts" to be the only biosynthesizers, and animals, unable to synthesize, as merely degrading the organic substances provided by the plants.

5. Prout's nutritional theories

Prout[30] objected to the theory of animalization by progressive increase of nitrogen content, as proposed by Fourcroy and Hallé, on the basis that if it were true, fasting animals would not have to respire. Starting from the composition of milk, in which fat, a sugar and casein are found, Prout concluded that the nutrients, as well as the substance of organisms, can be generally classified into three kinds of compound: saccharine, oleaginous

References p. 126

Plate 22. Jean Noël Hallé.

and albuminous, corresponding to our categories of carbohydrates, fats and proteins. Magendie[31] showed experimentally that none of the three categories is able, whatever the quantities given, to ensure adequate nutrition.

Prout characterized each of his three categories of nutrients in terms of elementary composition. Among the "albuminous", he distinguished albumin and gelatin, the latter containing less carbon; "oleaginous" were considered by him to be combinations of different amounts of water with ethylene ("olefiant gas"), all containing a large proportion of hydrogen; the "saccharine" class contained hydrogen and oxygen in the proportions of water, individual substances differing by the ratio of these elements to the amount of carbon. The different steps of "animalization" conceived by Hallé became unnecessary in Prout's scheme in which he proposed that the different classes of compounds in food were easily changed into corresponding compounds in fluids and solids of the body. He considered that proteins and fats did not need any basic changes of composition in assimilation, and he supposed that sugar is converted into one of the other two categories.

Prout had observed that minute quantities of foreign bodies (impurities) prevented the crystallization of dissolved substances. Their presence therefore changes the properties of these substances. This, in his opinion, was the result of a transformation similar to those that the chemical species underwent when present in the substance of the organisms. Prout believed that in digestion the elements of water are added to each class of nutrient, making them less stable, some of these elements being removed when they are converted into the body substance, the resulting water being exhaled from the lungs. Prout thought that, in the capillaries, a certain proportion of the carbon of the albumin is removed, the removal converting it into the gelatin of what we now call connective tissue, and this carbon being used in respiration. But many people opposed such concepts because they could not avoid the fact that several animals were herbivorous. How did such animals obtain the nitrogen of their tissues, as the vegetable substances, supposed to provide the matter of the animal body, contained a smaller overall proportion of nitrogen?

Macaire and Marcet[32], when endeavouring to solve this difficulty, determined, according to the habits of the "analysts", the proportions of nitrogen in whole blood and in chyle. Finding that the proportion of nitrogen in the chyle of a dog equals that in the chyle of a horse, and finding

References p. 126

Plate 23. William Prout.

that blood contains more nitrogen than the chyle, they developed a conjectural theory according to which it was claimed that animals acquired some nitrogen by respiration.

Boussingault's views[33] had a deeper impact. Performing elementary analyses of portions of food from a uniform diet, as well as from the milk, urine and excrements of a cow, he concluded that the fodder consumed affords enough nitrogen to account for the needs. This had been the view expressed by Magendie (1816) who had concluded that food is the source of the nitrogen of animal organisms[34]. Boussingault, in his treatise on *Économie Rurale*[35], insisted on the necessity of providing nitrogen in the diet of animals.

It is important here to note that agricultural interests played an important role in the development of metabolic theories during the 19th century. It is sometimes stated that agriculture remained a practice without a theory until the late 18th century. This ignores the fact that B. Palissy attempted to develop such a theory. He introduced the concept of the importance of certain "salts" as essential plant nutrients. The common belief that this work had been overlooked and had no influence on farming practices, overlooks the wide diffusion in England of the work of Plat, a disciple of Palissy[36]. The interest in the chemical basis of agriculture was actively revived by Lavoisier's circle and by his followers, including Liebig, who believed that a knowledge of the elementary composition of the components of plants and animals would provide an understanding of the basis of agricultural practice.

6. Mulder's theory

Elementary analyses were used by Mulder, from 1837 onwards, in a way that aroused great attention. In 1838, Mulder wrote a letter[37] to Liebig who, after having been a collaborator of Gay-Lussac, for several years had been developing methods of elementary analysis and had applied them to the analysis of a number of organic natural substances. His purpose was to determine the laws of chemical change in the compounds forming animal and vegetable substances in order to show how each atom of the substance taking part in the biological process was to be accounted for in the products of the reaction. In this letter, which was published by Liebig, Mulder reported the results of the elementary analyses of fibrin, egg albumin, serum "albumin" and wheat "albumin". He concluded that they all contained a

References p. 126

Plate 24. Gerhardus Johannes Mulder.

substance (radical) $C_{40}H_{62}N_{10}O_{12}$ (protein) and a few atoms of phosphorus and sulphur:

"Ich halte es nun für festgestellt, dass die Hauptmasse der animalischen Stoffe unmittelbar aus dem Pflanzenreich geliefert wird, und dass das Fibrin und Albumin [obviously animal as well as vegetable], bis auf 1 Atom Schwefel, gleiche Zusammensetzung haben", and "Um die analysirten Körper zu unterscheiden, habe ich die in dem Fibrin u.s.w. enthaltene tierische Materie *Protein* genannt; $\overline{Pr}$+SP ist also Fibrin; $\overline{Pr}+S_2P$ Albumin von Serum u.s.w." [($\overline{Pr}$ = varying number of $C_{40}H_{62}N_{10}O_{12}$ radicals)] (ref. 37)

Having, with his collaborators, analyzed many more albumin-like substances, he gave[38] a number of other examples: crystallin, 15 protein + SP; casein, 10 protein + S; "Pflanzenleim", 10 protein + S_2; fibrin, 10 protein + SP; serum albumin, 10 protein + S_2P.

The term "protein" designating a radical common to all vegetable and animal proteins, was introduced into science by Mulder in his letter to Liebig of 1838. The word was coined by Berzelius. This historical detail is reported by Mulder himself in his autobiography[39] in the following sentences:

"The egg-white like bodies (albumin) have occupied a considerable part of my life. All kinds of difficulties had to be surmounted, difficulties not met with other bodies; and whatever may have been said about this, one point is certain, that I have been the first who has shown (in 1838) that the meat is present in the bread and the cheese in the grass; that the whole organic kingdom is endowed with one and the same group, which is transferred from plants to animals and from one animal to another: one group, which is the first and foremost, and which I therefore still wish to call protein, a word derived from the Greek suggested to me by Berzelius." (translation of Westenbrink[40])

It appears that the idea that animal "protein" is formed in plants and transferred to animal substance without basic changes also originated from Berzelius[41].

References p. 126

REFERENCES

1 H. Guerlac, *Lavoisier, The Crucial Year, The Background and Origin of his First Experiments on Combustion in 1772*, Ithaca, 1961.
2 A. L. Lavoisier, *Mém. Acad. Roy. Sci.*, (1777) 182, 185.
3 A. L. Lavoisier and P. S. Laplace, *Mém. Acad. Roy. Sci.*, (1783) 48.
4 M. Berthelot, *La Révolution Chimique, Lavoisier*, Paris, 1902.
5 J. P. Partington, *History of Chemistry*, Vol. 3, London, 1962.
6 M. Foster, *Lectures on the History of Physiology during the 16th, 17th and 18th Century*, Cambridge, 1901.
7 F. L. Holmes, Introduction to Liebig's *Animal Chemistry*, facsimile ed., New York and London, 1964.
8 A. Seguin and A. L. Lavoisier, *Mém. Acad. Sci.*, (1789) [1793] 580; (1790) [1797] 606.
9 *Méthode de Nomenclature Chimique, Proposée par MM. De Morveau, Lavoisier, Bertholet (sic) et De Fourcroy*, Paris, 1787.
10 A. L. de Lavoisier, *Traité Élémentaire de Chimie*, 2 Vols., Paris, 1789.
11 E. Grimaux, *Lavoisier 1743–1794*, Paris, 1888.
12 E. Farber, *The Evolution of Chemistry*, New York, 1952.
13 N. le Fèvre, *Traité de Chimie*, 2 Vols., Paris, 1660.
14 N. Lemery, *Cours de Chymie*, Paris, 1675.
15 J. Dalton, *A New System of Chemical Philosophy*, Part I, Manchester, 1808.
16 S. C. Kapoor, *Proc. 10th Intern. Congr. Hist. Sci.*, 2 (1962) 879.
17 M. E. Chevreul, *Recherches Chimiques sur les Corps Gras d'Origine Organique*, Paris, 1823.
18 M. E. Chevreul, *Considérations Générales sur l'Analyse Organique et sur ses Applications*, Paris, 1824.
19 A. Fourcroy, *Éléments d'Histoire Naturelle et de Chimie*, 5th ed., Paris, 1793.
20 M. E. Chevreul, *De la Méthode a posteriori Expérimentale et de la Généralité de ses Applications*, Paris, 1870.
21 F. L. Hünefeld, *Physiologische Chemie des menschlichen Organismus*, 2 Vols., Leipzig, 1826.
22 F. V. Raspail, *Nouveau Système de Chimie Organique*, 2nd ed., 3 Vols., Paris, 1838.
23 J. Braun and O. Wallach (Eds.), *Briefwechsel zwischen J. Berzelius und F. Wöhler*, 2 Vols., Leipzig, 1901.
24 J. R. Partington, *History of Chemistry*, Vol. 4, London, 1964.
25 F. L. Holmes, *Isis*, 54 (1963) 50.
26 A. Fourcroy, *Ann. Chim. (Paris)*, 1 (1789) 254.
27 J. N. Hallé, *Ann. Chim. (Paris)*, 2 (1791) 158.
28 A. Richerand, *Nouveaux Éléments de Physiologie*, 4th ed., Paris, 1807.
29 L. Gmelin, *Z. Physiol.*, 3 (1829) 173.
30 W. Prout, *Chemistry, Meteorology and the Function of Digestion Considered with Reference to Natural Theology*, London, 1834.
31 F. Magendie, *Précis Élémentaire de Physiologie*, 4th ed., Brussels, 1834.
32 J. Macaire and F. Marcet, *Mém. Soc. Phys. et Hist. Nat. Genève*, 5 (1832) 223.
33 J. B. Boussingault, *Ann. Chim. Phys.*, 71 (1839) 113, 128.
34 F. Magendie, *Mém. Acad. Sci.*, (1816) 19 August.
35 J. B. Boussingault, *Économie Rurale Considérée dans ses Rapports avec la Chimie, la Physique et la Météorologie*, 2 Vols., Paris, 1843–1944.
36 A. G. Debus, *Arch. Intern. Hist. Sci.*, 21 (1968) 67.
37 G. J. Mulder, *Ann. Pharm.*, 28 (1838) 73.

38 G. J. Mulder, *Proeve eener Physiologische Scheikunde*, Rotterdam, 1843–1850. [German translation: *Versuch einer allgemeinen physiologischen Chemie*, Braunschweig, 1844–1851.]
39 G. J. Mulder, *Levensschets van G. J. Mulder door Hemzelven Geschreven en door Drie Zijner Vrienden Uitgegeven*, Rotterdam, 1881.
40 H. G. K. Westenbrink, *Clio Medica*, 1 (1966) 153.
41 H. B. Vickery, *Yale J. Biol. Med.*, 22 (1950) 387.

Chapter 6

The Nature of Alcoholic Fermentation, the "Theory of the Cells" and the Concept of the Cells as Units of Metabolism

1. Alcoholic fermentation

Fermentation, a natural process familiar to man in ancient times, is mentioned in the earliest protochemical writings, and in fact many ideas as well as much of the language of the alchemists is derived from fermentation.

For the alchemists, a ferment was an active substance the effect of which is to transform a passive substance, fermentable, into its own substance. If you take leaven and mix it with flour and water the mixture rises, and it can be seen that the leaven has changed the mixture into its own substance. Fire is a powerful ferment, and combustion is a fermentation, as the combustible becomes fire. The philosophers' stone, related to gold or silver, and transmuting base metals into gold or silver, was a ferment as it was capable of changing base metals into its own substance.

For the alchemists, alcohol was present before alcoholic fermentation took place, and they considered alcoholic fermentation as a separation process during which the impurities mixed with the alcohol were separated in the form of yeast. The evolution of gas that takes place in fermentation was distinguished by Francis Deleboë Sylvius in 1659 from other evolutions of gas such as that taking place when acids act on carbonate. The gas evolved, which had been called "gas vinorum" by Van Helmont (1648), was identified by MacBride (1764) with the fixed air of Black (CO_2). It was also recognized that no fermentation took place "quod non sit dulce", *i.e.*, that only sweet liquors could be fermented.

At the time of the phlogiston theory, Willis (1659) and after him Stahl (1697) explained fermentation as being a violent internal motion of the particles of the fermenting substance producing a simplification, with the

References p. 144

formation of new particles, among which were ethyl alcohol and CO_2*.

After he had introduced the procedure of elementary analysis, Lavoisier[3], following the principle of the presence of the same quantity of matter before and after a chemical operation, drew up an equation between the quantities of C, H and O in the sugar and in what he considered as the resulting substances, alcohol, CO_2 and acetic acid, according to which the products contained the whole matter of the sugar. A simple chemical change, the splitting of the sugar into alcohol and carbon dioxide, was thereby substituted for the mystery of alcoholic fermentation. With regard to the mechanism of fermentation, about which Lavoisier did not formulate any theory, Fabroni (1787) had expressed the view, soon to be widely spread, that fermentation was the result of the action on starch and sugar of the gluten derived from grain[4]. Thénard[5] opposed these views in 1803, but he remained uncertain about the nature of the "ferment" involved. Thénard differs, in his theory, from Lavoisier, as he ascribes the origin of some of the CO_2 to the carbon of the "ferment".

With respect to the "ferment", Gay-Lussac (1810) believed it was formed by the action of oxygen on the liquid[6]. This theory has its origin in the process of food preservation first introduced by Appert. In this process, the material was placed in bottles which were carefully closed and exposed for some time to the temperature of boiling water, a method still used nowadays by housewives.

When such a bottle was opened and contact with air was established, putrefaction began. Because an analysis of the air left in the sealed bottle of preserved food showed that all the oxygen had been absorbed, and because fermentation of preserved grape juice started at once when it was in contact with air, the theory according to which fermentation was set up by the action of oxygen on the fermentable material became standard. Gay-Lussac's opinion was that the "ferment" formed by the action of oxygen on the liquid was altered by heat and rendered incapable of acting, as occurs with brewer's yeast. The latter, on account of its insolubility, he concluded as being different from the regular agent of fermentation, considered by him as a "soluble ferment". In spite of his attributing a wrong composition to saccharose, Gay-Lussac's equation of the process rightly states that one molecule of glucose is split into two molecules of CO_2 and two molecules of ethyl alcohol. As nobody considered at the time

---

* On this ancient literature of fermentation, see Harden[1] and Haehn[2].

that yeast was the agent of fermentation, its presence in the fermentation mixtures remained puzzling.

In a letter written to the Royal Society in 1680, Leeuwenhoek reports the results of his microscopical observations on yeast. He describes small particles:

"some of these" he says "seemed to me to be quite round, others were irregular, and some exceeded the others in size, and seemed to consist of two, three or four of the aforesaid particles joined together. Others again consisted of six globules and these last formed a complete globule of yeast." (translation by Chapman[7])

As Keilin pointed out, there is no doubt that Leeuwenhoek failed to recognize the true nature of yeast. In 1826, Desmazières[9] recognized the elongated structure of yeast *Mycoderma cerevisiae*, but he did not suggest any relation with fermentation.

Cagniard-Latour, in 1838, published a paper[10] developing a communication he had presented on 12 June 1837 at the French Academy of Sciences (published in 1837)[11] and in which were summarized a number of notes he had read at the Academy or at the "Société Philomathique" since 1835. Cagniard-Latour's studies were exclusively microscopical and showed that fermentation depended on the presence of yeast and on a certain effect of its vegetation. Cagniard-Latour thought that fermentation took place in the liquid phase of the yeast suspension.

At about the same time, and independently of Cagniard-Latour, Schwann[12] attacked the problem from a different angle. Schwann started from his experiments proving that Infusoria do not result from a spontaneous generation and demonstrating the true nature of putrefaction. These experiments, important though they were as being the initial steps in microbiology, are outside our immediate interest, except in the measure that they influenced Schwann's views on fermentation. Schwann had demonstrated that an infusion of organic material, such as meat submitted to the temperature of boiling water, is kept free from putrefaction, even if air is passed through it, provided the air has been heated to a high temperature. By these experiments Schwann convinced himself that putrefaction was due to the presence of living germs in the air, and not, according to Gay-Lussac's theory, to an action of oxygen. Schwann answered any possible objection of a modification of oxygen by heating the air. He showed that a frog lives normally in previously heated air, and it is on the same lines that he turned to alcoholic fermentation as a phenomenon requiring oxygen, according to current views of the time. He described the

References p. 144

Plate 25. Friedrich Traugott Kützing.

morphology of yeast (fungus composed of the aggregation of globules) and observed that if the suspension of yeast is previously heated, and if heated air is passed through it, the yeast suspension does not show any sign of fermentation. It is therefore by chance, on the basis of experiments founded on a false current theory, that he recognized that fermentation is due to the development of a living organism he called "Zuckerpilz". He explained fermentation as

"the decomposition brought about by the sugar fungus removing from the sugar and a nitrogenous substance the materials necessary for its growth and nourishment, whilst the remaining elements of these compounds, which were not taken up by the plant, combined chiefly to form alcohol." (from ref. 12, translation by Harden[1])

Chronologically, as shown by his laboratory diaries (see Florkin[13]), the trail of research followed by Schwann started with experiments with heated air showing that putrefaction was the result of the generation of microorganisms, depending on the introduction of their germs (eggs) into the medium*. Recognizing the fact that heated air was still able to permit fermentation in a fermentable grape pressure juice previously heated, he was led to appreciate the living nature of yeast and linked fermentation to its multiplication. Observing that an addition of arsenic stops the process of alcoholic fermentation, he considered the latter as depending on the metabolism of microorganisms.

Cagniard-Latour's communications on fermentation by yeast started in 1835, and his extensive paper was presented at the Paris "Académie des Sciences" on 12 July 1837[11]. Schwann's results were first presented to the "Gesellschaft Naturforschender Freunde zu Berlin" in the first days of February 1837, and his more extensive report was published during the same year[12]. In July 1837, Kützing reported, at the "Versammlung des Naturhistorischen Vereins des Harzes" in Alexisbad, on microscopical studies showing the living nature of yeast and its participation in fermentation. He published an extensive paper on the same subject in 1837, but it has been proved that he had sent a paper on the subject to Poggendorff in December 1834 for publication and that Poggendorff had paid no attention to the manuscript and had lost it[14].

We can agree with the judgement passed by Harden[1] concerning the publications of Cagniard-Latour, Schwann and Kützing, that

"No less than three observers hit almost simultaneously upon the secret of fermentation and declared that yeast was a living organism." (Harden[1])

* "... ein Prinzip zur Infusorienbildung (Eier)"[13].

References p. 144

Plate 26. Theodor Schwann.

2. The cell theory

As noted by Canguilhem[15], in the field of biology the cellular theory prolonged the old debate over continuity and discontinuity. The search for a common structural principle of living creatures had preoccupied many scientists. In his biography of Virchow, Ackerknecht[16] distinguishes several successive aspects of this search for a common principle. In the 18th century the principle was the "fibre". According to this theory, which Ackerknecht calls "cellular theory No. 1", the development of fibres started from little globules. After this theory passed out of being, a new theory appeared in the school that Baker[17] has called the "globulist" theory (Ackerknecht's cellular theory No. 2) to which Oken, Meckel, Mirbel, Dutrochet, Purkynê, Valentin, Raspail, etc. belong. The concept of "globule" embraced a wide variety of supposed elementary units which were in reality nuclei, particles or optical illusions and sometimes cells. But a careful examination of the writings of the globulists shows that none of them, even if they occasionally figured a cell among their globules, can be regarded as having conceived of the organism as solely composed of cells, of modified cells or of products of cells. The latter notion only became the current view among botanists after 1830, when the perfecting of the microscope permitted the botanist Robert Brown to recognize the presence of the nucleus as the essential characteristic of plant cells, a view that was the basis of the cellular theory of plant structure as systematized by Schleiden[18].

In 1839, in his *Mikroskopische Untersuchungen*[19], Schwann formulated what Ackerknecht calls the 3rd cellular theory, which insists on the *common cellular origin of everything which lives.* By "cell" Schwann means a layer (*Zellenschichte*) around a nucleus. This layer could differentiate itself: covered by a membrane, growing hollow as a vacuole or fusing itself with other cells. The second part of the theory, "Schwann's error"[20], teaches that the cells form themselves around a nucleus, within a "blastema" of amorphous substance which can be the interior of a cell or a fluid part (intracellular or extracellular; endogenous or exogenous). This view he obtained from his friend Schleiden, and he accepted it on the basis of arguments concerning an alleged pre-existence of the nucleus in cartilage.

Ackerknecht's cellular theory No. 4, which has become standard nowadays, is that of Remak and of Virchow, the first part of which follows Schwann in acknowledging the cellular composition of organisms, with the

References p. 144

cell as the vital element, the bearer of all characteristics of life. The second part of this theory, expressed in the dictum "omnis cellula e cellula" contradicts Schwann's error when he admitted, after Schleiden, the formation of cells within a blastema from non-organized material. The cell theory was made widely known by Virchow (1877)[21] who justly emphasized the deep insight of Schwann's views on cell differentiation, which he discovered.

"What makes Schwann immortal is not the "cell theory" nor the doctrine of the origin of "hour glass forms"* from the "cytoblastema", but the demonstration of the *cellular origin of all tissues*. Later, we were able to build on this demonstration when, as a result of daily observation, the heretical thought of the *continual reproduction of cells within the individual* gradually germinated and grew." (ref. 21, translation by Rather[22])

Schwann's *Mikroskopische Untersuchungen* are composed of three parts. The first is devoted to a microscopical** study of the *chorda dorsalis* of frog's larvae. Studying this tissue, Schwann showed that it consists of polyhedral cells, which have in or on the internal surface of their walls a structure corresponding to the nucleus as already known in the plant cells. Young cells are formed within parent cells. Schwann also found the structure of cartilage to be in accord with what was known in the domain of plant structure. He believed that he had observed not only that the cartilage cells had a nucleus but that the cartilage cells originate by a formation of the nucleus around which the cell is formed. He therefore convinced himself that the cells of cartilage were derived from structures similar to plant cells (with nucleus, membrane and vacuole) or in his terminology that the "elementary parts" were derived from "cells" and in our language that cells are derived from embryonic cells. In the second part of the *Mikroskopische Untersuchungen*[19], Schwann brings in a demonstration of the same notion with regard to "elementary parts" of a much more specialized nature. He found that the varied forms of the "elementary parts" of tissues (our cells)—whether they belonged to epithelia, hooves, feathers, crystalline lenses, cartilage, bones, teeth, muscles, fatty tissues, elastic tissues or nervous tissues—are the products of cell differentiation.

The conclusion was that "elementary parts", though quite distinct in a

---

* At that time watches were in the shape of spheres, and the new cells were considered as being formed as excrescences similar in form to the glass of the watch.

** Schwann had the good fortune of being able to make his observations with the help of a microscope endowed with the improvements introduced by Amici and providing a correction to the spherical and chromatic aberrations of microscopical lenses.

physiological sense, develop according to the same laws in all organisms (either animals or plants). When traced backwards from their state of complete development, however differentiated, to their primary conditions, the elementary parts of organisms (our cells) appear as developments of cells (our embryonic cells). This unitary theory of *development* which he called "Zelltheorie" contrasts with previous theories of "cells and fibres".

In Chapter 11 we will return to the intellectual intentions that led Schwann to conceive the cell theory.

3. The "theory of the cells"

The third part of Schwann's *Mikroskopische Untersuchungen*, the "theory of the cells" is a physiological and biochemical treatise in which the term "metabolic phenomena" is coined and a number of metabolic aspects are considered.

When we consider their deep and lasting impact on the development of biochemistry, it is worthwhile to quote at length from Smith's translation of this third part, much less known than the two others and in which Schwann links respiration with cell metabolism and recognizes heat as derived from this cell metabolism. Schwann distinguishes two groups of phenomena, first those

"which relate to the combination of the molecules to form a cell, and which may be called the *plastic* phenomena of the cell; secondly, those which result from chemical changes either in the component particles of the cell itself, or in the surrounding cytoblastema; and which may be called *metabolic* phenomena (*τὸ μεταβολικον*, implying that which is liable to occasion or to suffer change)." (ref. 19, translation by Smith[23])

Whereas Schwann's theory of cell differentiation was a breakthrough, his theory of the origin of cells from a cytoblastema was, as noted above, erroneous, but nevertheless it is with reference to this that Schwann, by observation and experimentation on a suspension of yeast in a decoction of malt, reached his concept of the intracellular site of metabolism, of respiration as linked with cell metabolism. When he considered the decoction of malt as the cytoblastema of yeast, Schwann made no wrong assumption from the biochemical point of view, as the increase in the amount of yeast matter originates in the substances dissolved in their medium.

"The cytoblastema, in which the cells are formed, contains the elements of the materials of which the cell is composed, but in other combinations; it is not a mere solution of cell-material, but it contains only certain organic substances in solution. The cells, therefore, not only attract

References p. 144

materials from out of the cytoblastema, but they must have the faculty of producing chemical changes in its constituent particles. Besides which, all the parts of the cell itself may be chemically altered during the process of its vegetation. The unknown cause of all these phenomena, which we comprise under the term of metabolic phenomena of the cell, we will denominate the *metabolic power*. The next point which can be proved is that this power is an attribute of the cells themselves, and that the cytoblastema is passive under it. We may mention vinous fermentation as an instance of this. A decoction of malt will remain for a long time unchanged; but as soon as some yeast is added to it, which consists partly of entire fungi and partly of a number of single cells, the chemical change immediately ensues. Here the decoction of malt is the cytoblastema; the cells clearly exhibit activity, the cytoblastema, in this instance even a boiled fluid, being quite passive during the change. The same occurs when any simple cells, such as the spores of the lower plants, are sown in boiled substances.

"The universality of respiration is based entirely upon this fundamental condition to the metabolic phenomena of the cells. It is so important that, as we shall see further on, even the principal varieties of form in organized bodies are occasioned by this peculiarity of the metabolic process in the cells.

"Each cell is not capable of producing chemical changes in every organic substance contained in solution, but only in particular ones. The fungi of fermentation, for instance, effect no changes in any other solution than sugar; and the spores of certain plants do not become developed in all substances. In the same manner it is probable that each cell in the animal body converts only particular constituents of the blood. The metabolic power of the cells is arrested not only by powerful chemical actions, such as destroy organic substances in general, but also by matters which chemically are less uncongenial; for instance, concentrated solutions of neutral salts. Other substances, as arsenic, do so in less quantity. The metabolic phenomena may be altered in quality by other substances, both organic and inorganic and a change of this kind may result even from mechanical impressions on the cells." (ref. 19, translation by Smith[23])

4. Origin of Schwann's concept of intracellular metabolism

It is clear that the concepts of "metabolic power" and "metabolic phenomena" originated in Schwann's mind from his discovery of the chemical changes taking place inside the yeast cells and not, as accepted by Cagniard-Latour, in the liquid phase of the fermenting yeast suspension, and that by a prophetic induction more than by a demonstration, he extrapolated them to the multicellular forms when he formulated his "theory of the cells" in the third section of his book. A note at the bottom of pages 234–236 of the *Mikroskopische Untersuchungen* refers to an experiment he made at the time of his studies on cells★.

"I could not avoid bringing forward fermentation as an example [of the localization of the metabolic phenomena in the cells themselves], because it is the best known illustration of the operation of the cells, and the simplest representation of the process which is repeated in each cell of the living body. Those who do not as yet admit the theory of fermentation set forth by Cagniard-Latour, and myself, may take the development of any simple cells, especially of

★ See Appendix 3 (p. 331).

the spores as an example, and we will in the text draw no conclusion from fermentation which cannot be proved from the development of other simple cells which grow independently, particularly the spores of the inferior plants. We have conceivable proof that the fermentation-granules are fungi. Their form is that of fungi; in structure they, like them, consist of cells, many of which enclose other young cells. They grow, like fungi, by the shooting forth of new cells at their extremities; they propagate like them, partly by the separation of distinct cells, and partly by the generation of new cells within those already present, and the bursting of the parent cells. Now, that these fungi are the cause of fermentation, follows, first, from the constancy of their occurrence during the process; secondly, from the cessation of fermentation under any influence by which they are known to be destroyed, especially boiling heat, arsenate of potash, etc.; and thirdly, because the principle which excites the process of fermentation must be a substance which is again generated and increased by the process itself, a phenomenon which is met with only in living organisms. Neither do I see how any further proof can possibly be obtained otherwise than by chemical analysis, unless it can be proved that the carbonic acid and alcohol are formed only at the surface of the fungi. I have made a number of attempts to prove this, but they have not yet completely answered the purpose. A long test tube was filled with a weak solution of sugar, coloured of a delicate blue with litmus, and a very small quantity of yeast was added to it, so that fermentation might not begin until several hours afterwards, and the fungi, having thus previously settled at the bottom, the fluid might become clear. When the carbonic acid (which remained in solution) commenced to be formed, the reddening of the blue fluid actually began at the bottom of the tube. If at the beginning a rod was put into the tube, so that fungi might settle upon it also, the reddening began both at the bottom and upon the rod. This proves, at least, that an undissolved substance which is heavier than water gives rise to fermentation; and the experiment was next repeated on a small scale under the microscope, to see whether the reddening really proceeded from the fungi, but the colour was too pale to be distinguished, and when the fluid was coloured more deeply no fermentation ensued; meanwhile it is probable that a reagent upon carbonic acid may be found which will serve for microscopic observation, and not interrupt fermentation. The foregoing enquiry into the process by which organized bodies are formed may perhaps, however serve in some measure to recommend this theory of fermentation to the attention of chemists." (ref. 19, translated by Smith[23])

Both the cellular theory of fermentation and the concept of the cells as units of metabolism were eventually generally accepted, but only after a long series of polemics.

5. Yeast denied as an agent of alcoholic fermentation

The notion of the living nature of yeast was received with scepticism by many, and with scorn by Berzelius. But Turpin published in 1838 a confirmation of Cagniard-Latour's conclusions and a recognition of the living nature of yeast[24]. This was in contradiction with the sayings of the chemists. They were incensed by the audacity of the upholders of the theory of fermentation by yeast, put them in their black books and decided to make them pay for it.

At that time, the *Annalen der Pharmacie* was an international journal

References p. 144

edited by Liebig, Dumas and Graham. Dumas sent to this journal a German version of Turpin's paper. This was in contradiction with Liebig's theory of fermentation. Liebig, cock of the walk, who ruled over the whole fabric of the "biochemistry" of the time, had no positive argument to offer and he resorted to a most unpalatable and dishonest procedure*. He published immediately after the paper by Turpin[25] in the *Annalen der Pharmacie* an anonymous article[26] entitled "Das enträtselte Geheimnis der geistigen Gährung". This satirical text, the work of his bosom friend Wöhler, enriched by Liebig with some particularly ferocious details, presented a caricature of the works of Cagniard-Latour, Schwann and Kützing on the subject of the role of yeast in alcoholic fermentation. According to this facetious article, yeast suspended in water assumes the form of animal eggs which hatch with an unbelievable rapidity in a sugary solution. These animals, in the shape of an alembic, have neither teeth nor eyes but they have a stomach, an intestine and urinary organs. Immediately upon leaving the eggs, they throw themselves upon the sugar and devour it, and this is presented as penetrating their stomach to be digested, with the subsequent production of excrements. In a word, they eat sugar, expelling alcohol at the extremity of their digestive tube and carbonic acid through their urinary organs. Moreover, their bladder has the shape of a "Champagner-bouteille".

The acute Schwann, who in his twenties during his active career of discoverer (from the end of 1844 to the end of 1848) had piled up major discoveries**, was a timid man. Since the publication in 1837 of his paper on alcoholic fermentation he had been looked upon as the black sheep by the bigwigs of German chemistry. This did not deter him from accomplishing the work recorded in his *Mikroskopische Untersuchungen* and formulating the cell theory and the theory of the cells, but the publication of the facetious article and the merriment which it aroused in conformist circles was too much for him.

The facetious article on fermentation did not influence the career of Kützing who was a botanist, nor of Cagniard-Latour, an engineer who became famous for his many mechanical inventions, but the cruel treatment dealt out to Schwann by the beknighted scientific pontiffs of his time ridiculed him and made it impossible for him to pursue a scientific career in Prussia. After he failed in his candidature for a chair at the University of Bonn,

* A lesson for those who, like the alchemists, believe that the practice of laboratory work develops the ethical sense of the scientist. On this aspect see Florkin[27].

** See Florkin[13].

these disappointments drove him into exile: he became professor of anatomy at the Catholic University of Louvain in Belgium where the only requirement needed was to be a catholic, which he had always been and which, at the time of the *Kulturkampf*, had added to his difficulties in Prussia.

Schwann was not primarily an anatomist, and he lost much time and effort in the preparation of his anatomical lectures. His sensitive nature was deeply wounded, and the mainspring of enthusiasm and discovery was broken. The iron had entered into his candid soul and, still a youth who had not attained the thirties, he bowled out of a brilliant career as an investigator which had only lasted four years. Though he lived until a ripe old age, he made no further important addition to the advancement of knowledge[13].

6. Liebig's theory of fermentation

Shortly after the publication of the satirical paper on fermentation, Liebig published in the same journal another piece of "writing-table biochemistry" formulating the theory of alcoholic fermentation as being the result of instability of a substance occurring with the access of air to the nitrogenous substances of plant juices[28]. This theory was to enjoy a long popularity among chemists.

While Schwann's biological outlook was of the nature of a strict reductionism to chemical and physical laws, Liebig was a mind from quite a different mould. His theory of fermentation was related to his form of vitalism, as he exposes in his *Animal Chemistry*[29] and in a simpler and clearer way in his *Familiar Letters on Chemistry*[30]. His concept of the vital force was akin to the views of Barthez and of Bichat, and his recourse to oxygen derived from the views of his master Gay-Lussac. For Liebig, the substances composing living organisms, when relieved from the influence of the vital force, keep their properties only by inertia. The substance of the organisms in plants has been formed under the influence of the vital force dominating chemical forces, and which has supervised the directions of the attraction of the atoms, given movement to the molecules, controlled the form and the properties of their aggregates and at the same time opposed the antagonistic action of other forces: force of cohesion, heat, electric force, etc. In animals and plants, the formation of the tissues results from the control by the vital force opposing the forces which, outside the realm of the vital force, *i.e.* the solid parts of organisms, render impossible the

References p. 144

aggregation of atoms in higher combinations. These perturbing forces have been tamed by a control of chemical forces by vital force, as well as heat, for instance, favouring the formation of inorganic compounds by opposing other forces. As well as heat, vital force is able to modify the tendencies and power of chemical forces which it controls by orientation, stimulation, repression or complete inhibition. At death, the perturbing forces take over, released as they are from the control of vital force, and on contact with air or after the effect of weak chemical actions, a decomposition takes place with the advent of *fermentation*, *putrefaction* and *slow combustion*, which finally brings back the elements to the state they were in before their introduction into the frame of life phenomena. In order to bring a part of an animal or of a plant to the stage of fermentation or putrefaction, it is necessary that heat, a chemical action, or contact with oxygen or with hydrogen takes place. If it is in contact with air at a proper temperature, the must of grape juice ferments, and when the fermentation comes to an end the liquid clears and a yellow sediment, called yeast, is deposited. Similarly, in contact with air, milk coagulates and urine ferments. Putrefaction consists of a fermentation of organic nitrogenous or sulphurous compounds with a production of foul substances.

What is the cause of these phenomena? According to Liebig, they are the result of the complex composition of organic matters, their facility of change resulting from the weak degree of attraction which keeps the constituent atoms together. When they cease being protected by the vital force, these organic matters, in contact with the oxygen of the air, undergo, when it combines with them, an intermolecular movement which breaks the equilibrium of reciprocal attractions. In this way, a movement is also produced within the molecule and this movement, spreading to other molecules, keeps the fermentation going, unless the resistance of the other molecules is sufficient to neutralize the shock. On the other hand, it is clear that no substance comes from outside; in the formation of lactic acid from milk lactose for instance, the fermentation results from a rearrangement of the elements of prexisting molecules.

A non-nitrogenous constituent of animals such as starch, sugar or fat, does not ferment when in contact with oxygen. Other molecules containing nitrogen and sulphur, besides carbon, hydrogen and oxygen, when in contact with oxygen, are decomposed and become the real agents of fermentation, *i.e.* of the transformation of non-nitrogenous substances. Fermentation persists as long as the sugar and nitrogenous substances undergoing the transformation persist. If the access to oxygen is interrupted, both acts of transformation (of the nitrogenous agent and of the sugar) stop at once. For Liebig, fermentation and putrefaction are the first phases of what he conceives as a cycle of "slow combustion" progressively transforming the organic molecules into simpler substances. When the vital activity ceases to repress chemical forces, the constituents of the organisms return to the endless domain of these chemical forces.

It was necessary to wait for the work of Pasteur for justice to be done to Cagniard-Latour, Schwann and Kützing. The controversy between Liebig and Pasteur started in 1857 and was crowned by general recognition in 1872 of Pasteur's thesis. This controversy is aptly told by Harden[1]. It ended in the full recognition of Schwann's views on putrefaction and fermentation as well as of Pasteur's views on the other fermentations produced by bacteria. It struck a final blow to Liebig's theory of fermentation. This conclusion has sometimes, and indeed unfortunately, been qualified as "vitalist". To state that alcoholic fermentation is the result of the action of yeast had of course no more to do with "vitalism" or "vital force" than to conclude that tuberculosis is the result of the action of Koch's bacillus. Schwann was a determined antivitalist, whereas Liebig, on the contrary, was a confirmed vitalist*.

---

* On Liebig's vitalism, see also Lipman[31,32]

References p. 144

REFERENCES

1 A. Harden, *Alcoholic Fermentation*, London, 1911; 4th ed., 1932.
2 H. Haehn, *Biochemie der Gärungen*, Berlin, 1952.
3 A. L. Lavoisier, *Traité Élémentaire de Chymie*, Paris, 1789.
4 A. Fabroni, *Dell'arte di Fare il Vino*, Florence, 1787.
5 L. J. Thénard, *Ann. Chim. Phys.*, 46 (1803) 294.
6 L. J. Gay-Lussac, *Ann. Chim. Phys.*, 76 (1810) 245.
7 A. C. Chapman, *J. Inst. Brew.*, 28 [N.S.] (1931) 433.
8 D. Keilin, *The History of Cell Respiration and Cytochrome*, Cambridge, 1966.
9 Desmazières, *Ann. Sci. Nat.*, 10 (1826) 46.
10 Ch. Cagniard-Latour, *Ann. Chim. Phys.*, 78 (1838) 206.
11 Ch. Cagniard-Latour, *Compt. Rend.*, 4 (1837) 905.
12 T. Schwann, *Ann. Physik*, 41 (1837) 184.
13 M. Florkin, *Naissance et Déviation de la Théorie Cellulaire dans l'Oeuvre de Théodore Schwann*, Paris, 1960.
14 F. T. Kützing, in R. H. W. Müller and R. Zaunich (Eds.), *Aufzeichnungen und Erinnerungen*, Leipzig, 1960.
15 G. Canguilhem, *La Connaissance de la Vie*, Paris, 1952.
16 E. H. Ackerknecht, *Rudolf Virchow*, Madison, Wisc., 1953.
17 J. R. Baker, *Quart. J. Microscop. Sci.*, 89 (1948) 103.
18 M. J. Schleiden, *Contributions to Phytogenesis* (translation by H. Smith), London, 1847.
19 T. Schwann, *Mikroskopische Untersuchungen über die Übereinstimmung in der Struktur und dem Wachstum der Thiere und Pflanzen*, Berlin, 1839.
20 E. Mendelsohn, *Proc. 10th Intern. Congr. Hist. Sci., Ithaca, 1962*, Vol. 2, Paris, 1964, p. 967.
21 R. Virchow, *Arch. Pathol. Anat. Physiol.*, 70 (1877) 1.
22 J. L. Rather, *Bull. Hist. Med.*, 30 (1956) 537.
23 T. Schwann, *Microscopical Researches into the Accordance in the Structure and Growth of Animals and Plants* (translated by H. Smith), London, 1847.
24 E. Turpin, *Compt. Rend.*, 7 (1838) 369.
25 E. Turpin, *Ann. Pharm.*, 29 (1839) 100.
26 Anonymous, *Ann. Chem. Pharm.*, 29 (1839) 100.
27 M. Florkin, *Organon*, 6 (1969) 1.
28 J. von Liebig, *Ann. Pharm.*, 29 (1839) 250.
29 J. von Liebig, *Die Thierchemie oder die organische Chemie in ihrer Anwendung auf Physiologie und Pathologie*, Brunswick, 1842. [Translated by W. Gregory, *Animal Chemistry*, Cambridge, 1842.]
30 J. von Liebig, *Chemische Briefe*, Heidelberg, 1844 (translated by J. Gardner under the title of *Familiar Letters on Chemistry*, 1843; the translation appeared before the German text).
31 T. O. Lipman, *Isis*, 58 (1967) 167.
32 T. O. Lipman, *Bull. Hist. Med.*, 40 (1966) 511.

Chapter 7

The Rise and Fall of Liebig's Metabolic Theories

1. Agricultural chemistry

Liebig followed the path opened by Mulder. In 1841, he reported the isolation from plants of substances identical with the albumin, the fibrin and the casein of animals, obtaining the identity not only for elementary composition, but also for physical and chemical properties[1]. In the same paper, he formulated the principle according to which nitrogenous compounds alone contribute to the formation of animal substances, carbohydrates and fats playing no rôle in that respect.

"Organic chemistry", in the period considered, corresponded to our biochemistry, as stated by Liebig in his book on organic chemistry in its applications to agriculture and physiology[2]:

"The object of organic chemistry is to discover the chemical conditions which are essential to the life and perfect development of animals and vegetables, and, generally, to investigate all those processes of organic nature which are due to the operation of chemical laws". (translation by Playfair[2])

Liebig's book showed a deep interest in agricultural chemistry[3], well in line with Lavoisier's tendency and built on Lavoisier's principle of the conservation of matter. Liebig developed a theory that plants obtain carbon and nitrogen from the air, but obtain inorganic constituents from the soil, impoverishing it in potassium, chalk, sulphur, magnesium, calcium and iron. He concluded that these constituents must be returned to the soil in the form of fertilizers. Liebig was especially interested in increasing soil productivity by the use of inorganic fertilizers. At the time, Great Britain was the main consumer of fertilizers, in the form of guano imported from Peru, and of bones imported from Bavaria. The best method of compensating for the losses from the soil appeared to Liebig to be that practised in China: the integral restitution of excrements, either from man or animals.

References p. 162

Plate 27. Justus von Liebig.

In his book of 1840[2] Liebig, who was the most extreme of the "analysts" as defined above, and who based his reasonings on empirical formulas, formulated the "quantitative" relationships between the percentage compositions of various compounds which can be formed within plants, and the relative proportions of carbonic acid absorbed and oxygen released by these plants. These quantities he calculated by assuming that water and carbonic acid are the ultimate sources of the elements used in the synthesis and the oxygen is that left over according to the formulas of the respective compounds formed. If starch is being formed, for example:

"36 eq. carbonic acid and 30 eq. hydrogen
derived from 30 eq. water = starch
with the separation of 72 eq. oxygen

Or, if the organic product is oil of turpentine:

"36 eq. of carbonic acid and 24 eq. hydrogen
derived from 24 eq. water = oil of turpentine
with the separation of 84 eq. of oxygen

This method made specific and clear a relationship formerly perceived in a more general way between respiratory exchanges and the organic composition of living beings". (ref. 4)

In the same book, Liebig referred to nutrition processes of animals.

"The kidneys, liver, and lungs, are organs of excretion; the first separate from the body all those substances in which a large proportion of nitrogen is contained; the second, those with an excess of carbon; and the third, such as are composed principally of oxygen and hydrogen." (translation by Playfair[2])

Holmes[4] emphazises that in his book on agriculture, Liebig, in accordance with contemporary thought about "animalization" by increase of the nitrogen content of nutrients, states that the assimilation of food into blood and tissues "required chemical transformations of the nutrient substances into new compounds differing in their elementary compositions", while in 1843 he would claim that the nutrients (proteins), when they are eaten, already have the required elementary composition. Most of his biochemical views were the extension of results obtained *in vitro* to organisms. Only in one single case, immediately after the publication of his book on agricultural chemistry, in November 1840, did Liebig perform an experiment on organisms, repeating on a Hessian army company the experiment made by Boussingault on animals. From the average daily intake of carbon per man and the average amount excreted, he calculated the quantity burned in respiration daily, which was in line with the results of Boussingault[4].

References p. 162

Plate 28. Jean Baptiste Dumas.

2. Animal chemistry

Within a few months after the appearance of his *Agricultural Chemistry*[2], Liebig developed entirely new concepts about animal metabolism. By the end of 1840, he had concluded that the nitrogenous constituents provided by the vegetable nutrients were identical in elementary composition with the constituents of blood, and that sugar, starch, gum are not nutrients, serving only for the formation of fat and for respiration[5]. He had also reached the conclusion that bile is reabsorbed into the blood to provide carbon compounds for respiration. Adding the elements mentioned in the empirical formulas of urea and uric acid to those of the compounds he considered as characteristic of the bile, he obtained a sum corresponding to the empirical formulas of albumin or fibrin from blood. He thought he had found the right method of deciphering the complexities of metabolism, and he exploited the method extensively in the following months, in the course of intense activity.

In June 1841, Liebig thought he had isolated plant substances whose elementary formula was exactly identical with the animal substances albumin, fibrin, casein. He concluded that there are preformed plant substances providing each of the principal nitrogenous constituents of animals[6]. It must be pointed out that this theory differs from the theory of Mulder, who claimed that there was a single "protein", a radical from which the nitrogenous components of plants and animals derive by slight modifications. If the nitrogenous substances of animals are preformed in plants, it is clear why, as claimed by Liebig, fats and sugars play no part in the formation of animal matter.

In 1841 the *Essai de Statique Chimique des Êtres Organisés* by Dumas and Boussingault appeared[7,8]. Based on the final lecture (20 August 1841) in the course given by Dumas at the Paris Medical School, this book deals with the correlations between plant and animal metabolism. Animals and plants are contrasted according to the following summary (p. 151).

To this text, Dumas and Boussingault annexed a series of documents proving that all these opinions had been stated in public, or published, before 1839. The fact that Boussingault had demonstrated the fixation of nitrogen by certain plants is also emphasized. The *Essai* of Dumas and Boussingault was in a way a priority claim with regard to the opinions formulated by Liebig in his book of 1840[2]. It spurred Liebig on to accelerate the development of his studies of animal chemistry. In December 1841,

Plate 29. Jean Baptiste Boussingault.

Animal	*Plant*
Combustion apparatus	Reduction apparatus
Locomotor	Immobile
Burns carbon, hydrogen, ammonium	Reduces carbon, hydrogen, ammonium
Exhales carbonic acid, water, ammonium oxide, nitrogen	Fixes carbonic acid, water, ammonium oxide, nitrogen
Uses oxygen, neutral nitrogenous substances, fats, starches, sugars, gums	Produces oxygen, neutral nitrogenous substances, fats, starches, sugars, gums
Produces heat and electricity	Absorbs heat, extracts electricity
Restitutes its elements to air and earth	Borrows its elements from air or earth
Transforms organic matters into mineral matters	Transforms mineral matters into organic matters

he published in the *Annalen* what was going to become the first part of his *Animal Chemistry* with a note stating that he was compelled to accelerate the publication to avoid such plagiarism by the same authors who had already stolen his agricultural ideas. Dumas[9] answered immediately by developing the thesis that what he had published was either generally known, or taught by himself before Liebig had taught it publicly. Liebig answered with an open accusation of plagiarism[10]. We shall not dwell on this argument, as it is rather ridiculous to discuss priority in erroneous theories. Historians have passed judgement on these rivalries of the past by concluding that the theories had been put forward independently[11,12].

In 1842, Liebig[13] published his *Thierchemie (Animal Chemistry)*, which represents the culminating point of the metabolic studies following Lavoisier's doctrine and along the lines of the "analysts". It was the first widely read and circulated biochemical book, and in spite of the incredible amount of errors it contains, it exerted a considerable effect on the development of biochemistry. In the first part of the book (pp. 1–96) Liebig develops an extensive theory of metabolism.

Lavoisier's theory, according to which respiration is a form of combustion and is the only source of animal heat, had lost ground as soon as other sources of heat besides respiration, such as muscle contraction, were found in organisms.

Since 1810 the chemical theory of animal heat had been attacked by Brodie[14,15]. (For an analysis of his papers, see Goodfield[16].) From the experiments described in his first paper Brodie concluded that

"when the influence of the brain is cut off . . . no heat is generated notwithstanding the functions of respiration, and the circulation of the blood continues to be performed, and the usual changes in the appearance of the blood are produced in the lungs." (Brodie[14])

References p. 162

While he himself did not draw such conclusions, many at that time concluded that the heat production was not a chemical phenomenon, as postulated by Lavoisier and before him by the teachings of Black, but a manifestation of a "vital cause" operating through the nervous system.

In order to clear the issue, the French Academy of Science offered in 1824 a prize for an essay on animal heat. There were two answers, one from Despretz[17] and one from Dulong[18]. Both resorted to a simultaneous measure of heat production and respiratory exchanges. In the dissertations of both authors, the heat calculated from the amount of CO_2 produced and the heat of combustion of carbon was less than the heat measured in their water calorimeter. Despretz, who won the prize, had concluded that respiration was the main source of animal heat, and he accounted for the remaining part by the movements of the blood, by assimilation and by the friction of parts of the animal. Dulong was less conformist and he suggested that there must exist, besides respiration, another unknown source of heat.

In his *Animal Chemistry*, Liebig[13] states that the mutual action between the elements of the food and oxygen is the source of animal heat, and this statement became standard and was given to his credit, in spite of the experimental evidence which was against it at the time, in spite of the lack of evidence he could produce, and in spite of the fact that the principle had been expressed not only by Lavoisier but also by Dumas and Boussingault, who had already formulated in the 3rd edition of their *Essai* (1844)[7] the same unsubstantiated objection as proposed by Liebig, that the animals of Dulong and of Despretz produced excess heat because they were cooled.

The impact of Liebig's statement derived from his deserved fame as an organic chemist, and the fact was only demonstrated later, as we shall see in Chapter 11.

According to Liebig's theory, expounded in his books[2,13,19], animals obtain from their food the nitrogenous constituents necessary to replenish the composition of their blood and tissues (plastic nutrients). From their food, the herbivorous animals receive in addition non-nitrogenous carbonaceous molecules which they use for their respiration and for the production of heat in the blood (respiratory nutrients). Oxygen comes in contact with the tissues, but the vital force which resides in the tissues protects them from slow combustion. In a resting muscle, for instance, the affinity of the oxygen for the animal substance is completely inhibited by a corresponding amount of vital force. If a part of this vital force is, for instance, converted into movement, a corresponding amount of oxidation (slow combustion) takes place with a production of heat. The amount of vital force being reduced, the slow combustion takes place in an amount corresponding to the amount of vital force used up. The slow combustion (oxidation) decomposes the proteins into two parts. One, containing the greater proportion of nitrogen, is excreted by the kidney in the form of urea and uric acid. These compounds contain a proportion of carbon to nitrogen smaller than the substances forming the tissues. The other portion, the extra carbon, is in part submitted to combustion in the blood with a production of CO_2 and H_2O. The remaining part, carried to the liver, is secreted as "choleic acid", a compound containing a high percentage of carbon, and which was considered by Liebig as the main constituent of bile. When the bile enters the intestine, the "choleic acid" is reabsorbed into the blood where it is submitted to combustion. The decomposition products resulting from the release of slow combustion from the repression excreted by the vital force, are the only source of the combustion producing heat in the blood of carnivorous animals. This is why, when caged in zoos, such animals indulge in continual movement, to reduce by the utilization of vital force the amount of its repressing effect in the muscles, and consequently to liberate fuel for the combustion. As Liebig explicitly states,

"Our clothes are nothing but food equivalents: the more warmly we dress ourselves, the more, up to a certain point, our needs of eating diminish". (Liebig[19])

Since the nutrients supply directly or indirectly all the elements that are ultimately submitted to oxidation and combustion, the nourishment must be proportional in quantity to the oxygen absorbed in respiration. When insufficient oxygen is respired to oxidize the carbon and hydrogen available, they form fat which is stored.

For Liebig, the end-points of animal tissue metabolism, as different from combustion in the blood producing CO_2 and H_2O, were the nitrogenous components of urine and the carbonaceous components of bile. According to his theories, Liebig considered he was able to deduce the nature of the chemical changes taking place in the organism by comparing the elementary composition of tissue substances with the elementary composition of the products (nitrogenous compounds in urine and bile compounds). This method, which has sometimes been called "writing-desk biochemistry", he applied to a number of metabolic problems in the second section (The metamorphosis* of tissues) of the *Animal Chemistry*[13]. For carnivorous

* Metamorphosis, in Liebig's language, corresponds to metabolism.

References p. 162

animals, the method consists of adding to the elements composing the bile compound and to those of the organic constituents of urine, the elements of the protein-composing tissues, or what amounts to the same thing in his theory, to the elements of the proteins of blood, since the proteins of tissues are simply transferred as such from blood to solid parts. He balanced the equations with water, oxygen or ammonia, as seemed fitting.

As all the tissues are, in his theory, derived from albumin and fibrin, Liebig, in discussing the way in which this results from addition or subtraction of elements, reasons in the following way:

"If we now for example, look for an analytical expression of the composition of cellular tissue*, of the tissues yielding gelatine, of tendons, of hair, of horn, etc., in which the number of atoms of carbon is made invariably the same as in albumen and fibrine, we can then see, at the first glance, in what way the proportion of the other elements has been altered; but this includes all that physiology requires in order to obtain an insight into the true nature of the formative and nutritive processes in the animal body.

"From the researches of Mulder and Scherer we obtain the following empirical formulae:

Composition of organic tissues

Albumen	$C_{48}N_6H_{36}O_{14}+P+S$**
Fibrine	$C_{48}N_6H_{36}O_{14}+P+2S$
Caseine	$C_{48}N_6H_{36}O_{14}+S$
Gelatinous tissues, tendons	$C_{48}H_{75}H_{41}O_{18}$
Chondrine	$C_{48}H_6N_{40}O_{20}$
Hair, horn	$C_{48}N_7H_{39}O_{17}$
Arterial membrane	$C_{48}N_6H_{38}O_{16}$

"The composition of these formulae shows, that when proteine passes into chondrine (the substance of the cartilage of the ribs), the elements of water, with oxygen, have been added to it; while in the formation of the serous membranes, nitrogen also has entered into combination.

"If we represent the formula of proteine, $C_{48}N_6H_{36}O_{14}$ by Pr, then nitrogen, hydrogen and oxygen have been added to it in the form of known compounds, and in the following proportions, in forming the gelatinous tissues, hair, horn, arterial membrane, etc.

	Protein	*Ammonia*	*Water*	*Oxygen*
Fibrine, Albumen	Pr			
Arterial membrane	Pr		2 HO	
Chondrin	Pr		4 HO+2 O	
Hair, horn	Pr	$+NH_3$	3 O	
Gelatinous tissues	2 Pr	$+3\ NH_3$	HO+7 O	

* Our connective tissue.

** The quantities of sulphur and phosphorus here expressed by S and P are not equivalents, but only give the relative proportions of these two elements to each other, as found by analysis.

"From this general statement it appears, that all the tissues of the body contain, for the same amount of carbon, more oxygen than the constituents of blood. During their formation, oxygen, either from the atmosphere or from the elements of water, has been added to the elements of proteine. In hair and gelatinous membrane we observe, further, an excess of nitrogen and hydrogen, and that in the proportions to form ammonia." (translation of Gregory[13])

To give an example of the fantastic metabolic calculations piled up by Liebig's extreme "analysm" in the second section of his *Animal Chemistry*, let us turn to the way he considers the fate of caffeine or of theobromine:

"86. Without entering minutely into the medicinal action of caffeine (theine), it will surely appear a most striking fact, even if we were to deny its influence on the process of secretion, that this substance, with the addition of oxygen and the elements of water, can yield taurine, the nitrogenized compound peculiar to bile:

$$\begin{array}{lcr} \text{1 at. caffeine or theine} & = & C_8N_2H_5O_2 \\ \text{9 at. water} & = & H_9O_9 \\ \text{9 at. oxygen} & = & O_9 \\ \hline \end{array}$$

$$C_8N_2H_{14}O_{20} = 2 \text{ at. taurine} = 2\ (C_4NH_7O_{10})$$

"A similar relation exists in the case of the peculiar principle of asparagus and of althaea, asparagine; which also, by the addition of oxygen and the elements of water, yields the elements of taurine:

$$\begin{array}{lcr} \text{1 at. asparagine} & = & C_8N_2H_8O_6 \\ \text{6 at. water} & = & H_6O_6 \\ \text{8 at. oxygen} & = & O_8 \\ \hline \end{array}$$

$$C_8N_2H_{14}O_{20} = 2 \text{ at. taurine} = 2\ (C_4NH_7O_{10})$$

"The addition of the elements of water and of a certain quantity of oxygen to the elements of theobromine, the characteristic principle of the cacao bean (*Theobroma cacao*), yields the elements of taurine and urea, of taurine, carbonic acid, and ammonia, or of taurine and uric acid:

$$\begin{array}{lr} \text{1 at. theobromine} & C_{18}N_6H_{10}O_4 \\ \text{22 at. water} & H_{22}O_{22} \\ \text{16 at. oxygen} & O_{16} \\ \hline & C_{18}N_6H_{32}O_{42} \end{array} \quad = \quad \begin{array}{ll} \text{4 at. taurine} & C_{16}N_4H_{28}O_{40} \\ \text{1 at. urea} & C_2N_2H_4O_2 \\ \hline & C_{18}N_6H_{32}O_{43} \end{array}$$

or,

$$\begin{array}{lr} \text{1 at. theobromine} & C_{18}N_6H_{10}O_4 \\ \text{2 at. water} & H_{24}O_{24} \\ \text{16 at. oxygen} & O_{16} \\ \hline & C_{18}N_6H_{34}O_{44} \end{array} \quad = \quad \begin{array}{ll} \text{4 at. taurine} & C_{16}N_4H_{28}O_{40} \\ \text{2 at. carbon acid} & C_2\qquad O_4 \\ \text{2 at. ammonia} & N_2H_6 \\ \hline & C_{18}N_6H_{34}O_{44} \end{array}$$

References p. 162

or,

1 at. theobromine	$C_{18}N_6H_{10}O_4$		2 at. taurine	$C_8N_2H_{14}O_{20}$
8 at. water	H_8O_8	=	1 at. uric acid	$C_{10}N_4H_4O_6$
14 at. oxygen	O_{14}			
	$C_{18}N_6H_{18}O_{26}$			$C_{18}N_6H_{18}O_{26}$"

(translation of Gregory[13])

When considering the nitrogenous excreta of herbivora, Liebig writes:

"33. The urine of the herbivora contains no uric acid, but ammonia, urea, and hippuric or benzoic acid. By the addition of 9 atoms of oxygen to the empirical formula of their blood multiplied by 5, we obtain the elements of 6 at. of hippuric acid, 9 at. of urea, 3 at. of choleic acid, 3 at. of water, and 3 at. of ammonia; or, if we suppose 45 atoms of oxygen to be added to the blood during its metamorphosis, then we obtain 6 at. of benzoic acid, 13½ at. of urea, 3 at. of choleic acid, 15 at. of carbonic acid, and 12 at. of water.

		$5\,(C_{48}N_6H_{39}O_{15})+O_2$	$= C_{240}N_{30}H_{195}O_{84}$
	6 at. hippuric acid	$6\,(C_{18}NH_8O_5)$	$= C_{108}N_6H_{48}O_{30}$
	9 at. urea	$9\,(C_2N_2H_4O_2)$	$= C_{18}N_{18}H_{36}O_{18}$
=	3 at. choleic acid	$3\,(C_{38}NH_{33}O_{11})$	$= C_{114}N_3H_{99}O_{33}$
	3 at. ammonia	$3\,(NH_3)$	$= N_3H_9$
	3 at. water	$3\,(HO)$	$= H_3O_3$
	The sum is		$C_{240}N_{30}H_{195}O_{84}$

		$5(C_{48}N_6H_{39}O_{15})+O_{45}$	$= C_{240}N_{30}H_{195}O_{120}$
	6 at. benzoic acid	$6(C_{14}H_5O_3)$	$= C_{84}H_{30}O_{18}$
	27 at. urea	$27(CNH_2O)$	$= C_{27}N_{27}H_{54}O_{27}$
=	3 at. choleic acid	$3(C_{38}NH_{33}O_{11})$	$= C_{114}N_3H_{99}O_{33}$
	15 at. carbonic acid	$15\,(CO_2)$	$= C_{15}O_{30}$
	12 at. water	$12\,(HO)$	$= H_{12}O_{12}$
	The sum is		$C_{240}N_{30}H_{195}O_{129}$"

(translation of Gregory[13])

Summing up the metabolism of herbivora and in order to account for the "respiratory aliments" required by them according to his theory, Liebig adds the elements of starch to those of protein in order to balance the decomposition products:

"That is to say,—that if the elements of proteine and starch, oxygen and water being also present, undergo transformation together and mutually affect each other, we obtain, as the products of this metamorphosis, urea, choleic acid, ammonia, and carbonic acid, and besides these, no other product whatever.

"The elements of

5 at. proteine		9 at. choleic acid
15 at. starch	=	9 at. urea
12 at. water		3 at. ammonia
5 at. oxygen		60 at. carbonic acid

In detail

5 at. proteine	$5(C_{48}N_6H_{36}O_{14})$	$= C_{240}N_{30}H_{180}O_{70}$
15 at. starch	$15(C_{12}H_{10}O_{10})$	$= C_{180}H_{150}O_{150}$
12 at. water	$12(HO)$	$= H_{12}O_{12}$
5 at. oxygen		$= O_5$
The sum is		$C_{420}N_{30}H_{342}O_{237}$

and

9 at. choleic acid	$9(C_{38}NH_{33}O_{11})$	$= C_{342}N_9H_{297}O_{99}$
9 at. urea	$9(C_2N_2H_4O_2)$	$= C_{18}N_{18}H_{36}O_{18}$
3 at. ammonia	$3(NH_3)$	$= N_3H_9$
60 at. carbonic acid	$60(CO_2)$	$= C_{60}O_{120}$
The sum is		$C_{420}N_{30}H_{342}O_{237}$

"The transformation of the compounds of proteine present in the body is effected by means of the oxygen conveyed by the arterial blood, and if the elements of starch, rendered soluble in the stomach, and thus carried to every part, enter into the newly formed compounds, we have the chief constituents of the animal secretions and excretions; carbonic acid, the excretion of the lungs, urea and carbonate of ammonia, excreted by the kidneys, and choleic acid, secreted by the liver." (translation of Gregory[13])

Animal Chemistry, as we have said, was a most influential book. It embodies a synthesis of ideas scattered among Lavoisier's successors, the "analysts", who based their metabolic theories on the mere consideration of empirical formulas. As Holmes with great insight remarks,

"The view that the basic constituent compounds of animals are formed in plants and chiefly decomposed in animals by gradual oxidations reflects ideas such as those of Gmelin, rationalized by means of the identities in composition which Mulder found. The basic classification of sugars, fats and nitrogenous substances came from Prout, but Liebig assigned to each type more definite functions within the animal economy. The realization that nutrition must supply the elements which are oxidized in respiration, and therefore that the two processes must be proportional, originated with Lavoisier, although Liebig worked out many implications (not all correct) of that principle. Liebig's over-all scheme of nutrition was far more comprehensive than those of Fourcroy, Berthollet and Hallé, but his aim was the same as theirs. Each tried to show how the composition of the body tissue substances is maintained through the balanced addition and removal of their constituent elements via the food, respired gases and excrements." (ref. 20)

The period we consider is characterized by the transfer of the old concept of dynamic permanence from the body to molecular units. The view is clearly expressed by Liebig himself:

"The carbon and the carbonic acid given off, with that of the urine; the nitrogen of the urine, and the hydrogen given off as ammonia and water; these elements, taken together, must be exactly equal in weight to the carbon, nitrogen, and hydrogen of the metamorphosed tissues, and since these last are exactly replaced by the food, to the carbon, nitrogen and hydrogen of the food. Were this not the case, the weight of the animal could not possibly remain unchanged." (translation of Gregory[13])

References p. 162

"Whereas Hallé and Bertholet could only speak vaguely of qualitative increases or decreases in the proportions of elements, Liebig could speculate with chemical equations." (Holmes[20])

These equations had a tremendous appeal to the authors of handbooks, such as Marchand[21] (1844) or Valentin (1844)[22]. Valentin even calculated for man the formula of the "perspiration material" by subtracting the total of the elements in the excreta from the total of the elements in the food.

But the applause was far from being universal. Berzelius, to whom Liebig had dedicated his book, predicted that "after a few years surely none of these calculations will be retained in the science" (ref. 23), a sound judgement but one that stung Liebig bitterly. In a private letter to Wöhler, Berzelius[24] expressed his opinion in these terms: "Mein Gott, welche Radoterie!"

3. Liebig's followers

An analysis of the changes introduced in the second and third editions of *Animal Chemistry* is to be found in the introduction of Holmes[4] to the reprinted edition of the book. Holmes devotes a section of this introduction to Liebig's followers. As he points out, the aspect of Liebig's work that had the greatest attraction for his followers was the hope of measuring metabolism by the determination of rinary nitrogen. Frerichs[25] tried to determine the basic rate of tissue decomposition necessary to maintain the activity of an animal, and to reach this goal he measured the rate of urea secretion in fasting animals. He found this to be a small fraction of the excretion of an animal on a nitrogenous diet, and he logically concluded that the supplementary nitrogenous constituents beyond those needed for the basic rate were oxidized in the blood, just as sugars and fats are, according to Liebig. Frerichs' views, therefore, were in disagreement with those of Liebig for whom all nitrogenous nutrients must become part of organized tissues before being subject to the process of decomposition. The name of "Luxus consumption" was given to this supposed direct oxidation of nitrogenous nutrients (Bidder and Schmidt[26]). An objection was raised to this concept by Bischoff[27], since he observed that when the amount of protein is increased by stages in the food, the additions do not entirely appear in the urine, a portion of the added nitrogenous nutrient being retained as an addition to bodily substance. But when he calculated the amount of nitrogen excreted in urea and compared it with the amount in the food, he found that more than one-third of the food nitrogen remained unaccounted for. In 1857, Bischoff and Voit accomplished experiments in which urea nitrogen and food nitrogen balanced. They decided that, in their previous experiments and in those of Bischoff, they had falsely assumed that there is no change in composition when there is no change in weight. This restored their confidence in Liebig's views of the measure of muscle metabolism through urea measurements.

Bischoff and Voit[28] published in 1860 an imposing book in which a large number of experiments and observations were collected. Urea output was determined in fasting dogs, in dogs receiving gradually increased amounts of purely nitrogenous nutrients, or mixed diets of proteins and fat in which one was maintained constant and the other varied, or in which carbohydrates were substituted for fats. If *Animal Chemistry*[13] had been the culminating point of the post-Lavoisier period, Bischoff and Voit's book was its swan song. By this collection of an immense number of experiments, a confirmation was thought to be afforded of the keystone of Liebig's creed, that the sole source of muscular work was the decomposition of nitrogenous tissue substance. The authors maintain Liebig's view according to which all nitrogenous substances of the food form the organic constituents of blood, then the substance of tissues, where they are decomposed to provide muscular work, heat and urea. Of course, the fact that an increase of nitrogen in the diet, without any increase of muscular activity, resulted in an increase in urinary urea, posed a problem. But they pushed this aside by assuming that in these conditions there is an increase in the blood plasma volume, and that consequently there is a need of a greater amount of internal muscular work to circulate the blood. Holmes[4] briefly summarizes the theory as follows:

"Bischoff and Voit explained all their results by means of a theory that the rate of nitrogenous tissue decomposition is controlled by a kind of complex mass action effect. It is proportional respectively to the mass of tissue, to the amount of available oxygen, and to the amount of nutrients reaching the tissues, or the blood plasma volume. They found that additions of fat or carbohydrate to the diet substantially reduced the amount of nitrogenous nutrients necessary to maintain the body weight, but that no matter how much they added, they could not cause the metamorphosis of nitrogenous substances to fall below a certain level. In accordance with Liebig's theories, Bischoff and Voit interpreted this result to mean that fats or sugars cannot substitute for nitrogenous substances in the metabolism of tissues, but that by competing for the oxygen available in the blood, these non-nitrogenous substances could decrease the rate at which the tissue substances are oxidized to produce heat." (ref. 4)

According to the theory, when the nitrogenous substances of muscle are degraded in the course of muscular work there should result an increase of urea excretion when muscular activity takes place. Voit[29] did not observe this increase. Instead of abandoning Liebig's theory, Voit elaborated a complicated explanation in which Liebig's theory received a last stirring of the dry bones.

References p. 162

Plate 30. Carl von Voit.

In 1860, the aspect of Liebig's theory that appeared most difficult to accept was the concept of the production of muscular energy at one place and the production of heat at another. This did not fit in with the views on the conservation of energy that were taking shape at that time. The notion according to which the degradation of proteins was correlated with diet, and not with exercise, was also becoming generally accepted as well as the fact that, when muscular activity increased, there was an increase in the respiratory rate rather than an increase in the excretion of urea.

4. Rupture with Liebig's views

The rupture with Liebig's views was a process inaugurated by Traube (1861)[30], who believed that the oxidation of non-nitrogenous substances must play a role in the production of mechanical energy. Traube compared the heat of combustion of foodstuffs with the heat equivalent of muscular work, and was able to show that the portion of the diet that contributes energy for muscular activity was in no way as small, in relation to the portion serving for the maintenance of temperature (resting conditions), as had been believed.

But the decisive blow came, in 1866, from Fick and Wislicenus[31]. They climbed a mountain, calculated the work done and, as it was possible to do at that time, compared it with the energy corresponding to the catabolism of the amount of protein equivalent to the urea they had excreted. In spite of the crude nature of the calculations involved, it was shown that the work performed could not be provided by the proteins degraded.

In order to accomplish more accurate calculations, Frankland[32] made extensive calorimetric determinations of the heat released in the combustion of nutrient compounds, and this made the conclusions of Fick and Wislicenus even more decisive.

Now, Liebig, who had been silent for 10 years on the subject of biochemistry, published in 1870 a book[33] answering the criticisms raised against his theories. He recognized that he had been wrong in considering urea excretion as a measure of muscular work and tissue metabolism (Stoffwechsel, Metamorphose) but continued to pull a very long bow and retained the idea according to which nitrogenous "metamorphosis" is the source of muscular work. He even declared that some of the experiments accomplished by Voit on the metabolism of fat were of no interest. Voit exploded, and his reply (1870) came without delay[34]. He made no bones

References p. 162

about razing to the ground the metabolic system built up by Liebig, whose theory thus came to grief.

REFERENCES

1 J. von Liebig, *Ann. Chim. und Phys.*, 39 (1841) 129.
2 J. von Liebig, *Die organische Chemie und ihre Anwendung auf Agricultur und Physiologie*, Brunswick, 1840. (English translation by L. Playfair, 1840)
3 J. L. Dumas, *Rev. Hist. Sci.*, 18 (1965) 73.
4 F. L. Holmes, Introduction to *Animal Chemistry* by J. von Liebig, New York and London (Reprint), 1964.
5 J. von Liebig, *Ann. Chem.*, 41 (1842) 352.
6 J. von Liebig, *Ann. Chem.*, 39 (1841) 129.
7 J. B. A. Dumas and J. B. Y. D. Boussingault, *Essai de Statique Chimique des Êtres Organisés*, Paris, 1841. [An anonymous English translation appeared in 1844: *The Chemical and Physiological Balance of Organic Nature: an Essay*, the last word in the title misprinted as Cssay, as noted by Partington[8]]
8 J. R. Partington, *History of Chemistry*, Vol. 4, London, 1964.
9 J. Dumas, *Ann. Chim. et Phys.*, 4 (1842) 115.
10 J. von Liebig, *Ann. Chim.*, 41 (1842) 351.
11 A. W. Hofmann, *Zur Erinnerung an vorangegangene Freunde*, 3 Vols., Brunswick, 1888.
12 J. Volhard, *Justus von Liebig*, 2 Vols., Leipzig, 1909.
13 J. von Liebig, *Die Thierchemie oder die organische Chemie in ihrer Anwendung auf Physiologie und Pathologie*, Brunswick, 1842. [English translation by W. Gregory, *Animal Chemistry or Organic Chemistry in its Application to Physiology and Pathology*, 1842]
14 B. C. Brodie, *Philos. Trans. Roy. Soc.*, (1811) 36.
15 B. C. Brodie, *Philos. Trans. Roy. Soc.*, (1812) 378.
16 J. Goodfield, *The Growth of Scientific Physiology*, London, 1960.
17 C. Despretz, *Ann. Chim. Phys.*, 26 (1824) 337.
18 P. L. Dulong, *Ann. Chim. Phys.*, 1 (1841) 360 [The paper had been read at the French Academy in 1822].
19 J. von Liebig, *Chemische Briefe*, Heidelberg, 1844.
20 F. L. Holmes, *Isis*, 54 (1963) 50.
21 R. F. Marchand, *Lehrbuch der physiologischen Chemie*, Berlin, 1844.
22 G. Valentin, *Lehrbuch der Physiologie des Menschen*, Vol. 1, Braunschweig, 1844.
23 J. J. Berzelius, *Jahresberichte*, 23 (1844) 582.
24 J. Braun and O. Wallach (Eds.), *Briefwechsel zwischen J. Berzelius und F. Wöhler*, Vol. 2, Leipzig, 1901, p. 411.
25 F. Th. Frerichs, *Arch. anat. Physiol. wiss. Med.*, (1848) 469.
26 F. Bidder and C. Schmidt, *Die Verdauungssäfte und der Stoffwechsel*, Mitau, 1852.
27 T. L. W. Bischoff, *Der Harnstoff als Maas des Stoffwechsels*, Giessen, 1853.
28 T. L. W. Bischoff and C. Voit, *Die Gesetze der Ernährung des Fleischfressers*, Leipzig, 1860.
29 C. Voit, *Untersuchungen über den Einfluss des Kochsalzes, des Kaffee's und der Muskelbewegung auf dem Stoffwechsel*, Munich, 1860.
30 M. Traube, *Arch. Pathol. Anat.*, 21 (1861) 386.
31 A. Fick and J. Wislicenus, *Phil. Mag.*, 31 (1866) 485.
32 E. Frankland, *Phil. Mag.*, 32 (1866) 182.
33 J. von Liebig, *Über Gährung, über Quelle der Muskelkraft und Ernährung*, Leipzig, 1870.
34 C. Voit, *Z. Biol.*, 6 (1870) 303, 399.

Chapter 8

The Intracellular Location of Metabolic Changes

1. Lagrange and the theory of the location of respiration in the blood

Referring to the mechanism of respiration (*i.e.* what was considered as the conversion of O_2 into CO_2), Lavoisier and Laplace (1780) had developed a concept according to which respiration is a combustion, produced within the bronchi, without evolving perceptible light, of carbon and hydrogen exuded from the blood, by the oxygen of air. Lagrange claimed that combustion would not take place only in the lungs on the basis that the tissue would be destroyed by the rise in temperature, an argument which was later shown to be wrong by Berthelot[1] since the amount of heat produced would not warm the lungs appreciably. Lagrange suggested that oxidation takes place in the blood, and this theory became universally accepted. There is no publication by Lagrange on this subject. His opinion is related in a paper by a disciple of Lavoisier, Hassenfratz[2],★ who had become Lagrange's secretary.

As recorded by Hassenfratz, Lagrange supposed that in the lungs, the blood took up oxygen from the air (by dissolution, as the carrier function of haemoglobin was still unknown) and that it carried this oxygen into the arteries and from there to the veins, and that

"in the course of the journey of the blood the oxygen little by little quitted the condition of dissolution in order to combine in part with the carbon and in part with the hydrogen of the blood, and so to form carbonic acid and water, which are set free from the venous blood as soon as this leaves the right side of the heart to enter the lungs." (ref. 2, translation of Foster[4])

Hassenfratz adds to this statement of Lagrange's theory the results of a series of experiments of his own, but these experiments of Hassenfratz do

★ In 1797 a dissertation by Ruiz de Luzuriaga[3] appeared, which had been presented in 1790 at the Real Academia Medica of Madrid and in which he independently reached the same conclusion as Lagrange.

References p. 171

Plate 31. Joseph Louis Lagrange.

not amount to much as they were, in the main, repetitions of Priestley's earlier experiments.

The views of Lagrange seemed to be confirmed by the experiments of Magnus[5] to which we shall return in the study of gas exchanges in the accomplishment of the respiratory functions of the blood. In our present concern, Magnus[5] came to the conclusion that carbonic acid was not produced in the lungs, but in the blood capillaries, *i.e.* in blood. As the production of heat was linked with the production of carbonic acid, it was generally believed that this also took place in the blood of capillaries.

It is sometimes stated in historical texts that Magnus recognized the intracellular nature of respiration, which is contrary to the evidence. Magnus believed he had shown that the phenomenon took place in the blood itself. The steps in the transition between the theory of combustion in the blood and the theory of combustion in tissues have been clearly grasped by Scheer[6], by Mendelsohn[7] and by Keilin[8].

In spite of the wide recognition of the views of Lagrange made known by Hassenfratz, a number of authors still accepted in its original form the view formulated by Lavoisier and Laplace. Other authors suggested that the secretion of combustibles in the cavity of the bronchi could be dispensed with, and they proposed that while still taking place in the lung, the combustion was located in the blood of the pulmonary vessels. It was the "acute Spallanzani"[4] who refuted the secretion theory of Lavoisier by showing that the tissues respire, consuming oxygen and producing carbonic acid, and that snails placed in nitrogen or hydrogen give out carbonic acid as in air. But this contribution remained unknown until 1807, after Spallanzani's death (1803), when they were published by Senebier[9]. In spite of this publication, these ideas failed to produce any effect, though they were presented again in a forceful way by Edwards[10] in 1823.

It was Liebig's son, G. von Liebig[11], who, in a study of muscle respiration, reopened the problem in 1850 on the line formulated by Schwann (see Chapter 6). He showed that a frog muscle, separated from the body, takes up oxygen and liberates carbon dioxide. This was confirmed by Traube[12].

In his *Leçons sur les Propriétés Physiologiques et les Altérations Pathologiques de Liquides de l'Organisme*[13] Bernard reported measurements of respiration *in vitro* for liver, kidney, muscle or brain. In spite of this, he reverted to the theory of Lagrange, according to which respiration and the production of CO_2 take place in the blood. The aspect that worried him

References p. 171

Plate 32. Eduard Pflüger.

and led him astray was the difference in colour between arterial and venous blood. After many physiological experiments, he concluded that the darkening was due to an increase in the ratio CO_2/O_2.

Claude Bernard also believed at that time (1859) that there was no exchange of gas taking place between blood and tissues, but only an exchange of liquids. He expressed his view as follows.

"But if it is true, and we are much inclined to admit it is so, that the venous blood owes its black coloration to carbonic acid, we must recognize that the modification by which its oxygen could be transformed into carbonic acid can be brought about directly in it and not necessarily by immediate contact with tissues." (ref. 13, translation of Keilin[8])

Further, in the same book (p. 342) he writes:

"It is infinitely probable that the carbonic acid of venous blood results from an oxidation which is brought about within the red blood corpuscle itself. When the blood traverses the capillaries, there will be between it and the tissues not an exchange of gas but perhaps one of liquids. Following the new condition which such an exchange would create, the oxygen of the red blood corpuscle would be partly used for the oxidation of the carbon of the corpuscle itself." (ref. 13, translation of Keilin[8])

In 1862, Hoppe[14] (later called Hoppe-Seyler) described the absorption spectrum of oxyhaemoglobin, and two years later Stokes[15] showed its reversible reduction and described the absorption spectrum of the purple reduced haemoglobin. Paul Bert's book, published in 1870[16], gave further support to the notion of some respiration taking place in all the tissues, in spite of his contention that its major part was located in the blood, where from the combustion of the foodstuffs brought by the chyle, the major part of the CO_2 originated according to him. Influenced by the experiments of Schönbein[17-20] and of Schmidt[21], who maintained that the oxidizing properties of the blood in the reaction of guaiacum oxidation was a result of the presence of ozone, Bert still thought that a part of the oxygen of blood was in the form of ozone.

The method of perfusing isolated organs with defibrinated blood, and improved methods of gas analysis were exploited in Ludwig's laboratory. Ludwig's group concluded in favour of the CO_2 formation at the level of blood and suggested that oxidation or combustion predominates in the blood and not in the tissues[8]. Hammarsten[22] concluded that the amounts of oxygen yielded to tissues from blood were too small to account for the respiration on the site of tissues. The persistence in the belief in ozone in blood was due to the observation of the guaiacum reaction, the investigators being unaware that their reagents contained small amounts of H_2O_2 and

Plate 33. Hermann von Helmholtz.

that haematin derivatives or haemoglobin, or even salts, would give the reaction.

2. Intracellular respiration

A violent polemical discussion was started when Pflüger opposed Ludwig and demonstrated his errors. Pflüger[23,24] clearly grasped the significance of intracellular respiration and was the first to recognize the true function of blood in the respiration of organisms. The importance of Pflüger's contribution has been clearly stated by Keilin[8] in the following text.

"In his two general papers on the subject, published in 1872 and 1875, he clearly formulated the problem of the localization of respiratory activity in organisms and established once and for all that respiration is an intracellular process. This, he found, applied equally well to both plants and animals, including among the latter all forms from the protozoa to higher animals. However, in no group of the animal kingdom could he find this generalization so well illustrated as among insects, in which tracheal tubes filled with air penetrate directly into the tissues and cells and supply them with oxygen. He considered that in insects nature offers us invaluable experimental material, the importance of which can hardly be overestimated. (Pflüger, 1875, pp. 272, 273)[24]

"With regard to the part played by the blood in the respiration of higher organisms, Pflüger and his group (Wolffberg[25], 1872; Strassburg[26], 1872; Finkler[27], 1875; and others) clearly demonstrated by highly quantitative experimental methods that it is confined to carrying oxygen to, and carbon dioxide from, the tissues. In contradiction to Ludwig, Pflüger found that it is the cells and tissues which regulate the respiratory activity of an organism, and not the blood or circulation; haemoglobin can therefore be regarded only as a large-capacity, convenient, oxygen carrier (1875, ref. 24, p. 275). These views were further corroborated by two important experiments carried out in Pflüger's laboratory. First, Finkler[27] (1875) showed that bleeding or the slowing down of the circulation rate in a dog produced no effects on its oxygen uptake, and later Oertmann[28] (1877) demonstrated that a frog in which the blood was replaced by saline solution had the same respiratory activity as a normal frog. Finally, after careful criticism, Pflüger dismissed all the evidence put forward by Schmidt[29] and others that blood contains ozone, which imparts to it certain oxidizing properties. Not only did Pflüger reject this view, but he also demonstrated that cellular combustion is to a very great extent independent of the partial pressure of ordinary "inactive" oxygen. Moreover, higher partial pressures of ordinary oxygen do not increase the respiratory activity of an organism; on the contrary, it had been shown by Bert[16] that animals exposed to oxygen pressures above six atmospheres die in convulsions and that this is not due to the purely mechanical effect of the higher pressure, but specifically to the increased tension of the oxygen." (ref. 8)

The notion of the cellular location of respiration and of the regulation of respiration by cells is clearly stated by Pflüger over and over again. For instance, in his general paper of 1875[24], he writes:

"it is the living cell which regulates the magnitude of the oxygen consumption." (translation of Keilin[8])

Previously, a member of Schwann's circle in Berlin, Helmholtz, demonstrated experimentally in 1847 (publication in 1848[30]) the generation of heat by the twitch of isolated muscle.

Pflüger's demonstration of tissue respiration was accepted by Bernard who popularized it in his book on animal heat (1876)[31]. It was generally believed after this that the blood serves only as a carrier of oxygen to, and of carbon dioxide from, the tissues and has, as Hoppe-Seyler (1890)[32] re-asserted, no significant proper respiration.

As shown above, there was for many years a resistance to the concept of intracellular respiration. This must be related to the prevalence of physiological concepts at the time. The physiologists were used to dealing with whole organisms and they were, in the field of respiration, used to finding their evidence in experiments on blood gases. Their main interest was in the control of respiration rate and, as pointed out by Culotta[33], it was only when Pflüger placed his demonstrations in the framework of this control, that they met general acceptance.

After the demonstration by Pflüger of the intracellular nature of respiration and after the demonstration by Helmholtz of the production of heat by an isolated muscle, general assent was given to the views first expressed by Schwann in 1839, when on the basis of arguments taken from the field of microbiology, which carried no weight with the physiologists, he recognized the "elementary parts" which we call the cells, as being the site of respiration and heat production, and not only as units of structure but also as units of metabolism.

REFERENCES

1 M. Berthelot, *Compt. Rend.*, 77 (1873) 1063.
2 J. Hassenfratz, *Ann. Chim. et Phys.*, 9 (1791) 266.
3 I. M. Ruiz de Luzuriaga, *Memor. Real Acad. Medica*, 1 (1797) 1.
4 M. Foster, *Lectures on the History of Physiology During the 16th, 17th and 18th Centuries*, Cambridge, 1901.
5 G. Magnus, *Ann. Phys. Chem.*, 40 (1837) 583.
6 B. Scheer, *Ann. Sci.*, 4 (1939) 295.
7 E. Mendelsohn, *Heat and Life*, Cambridge, 1964.
8 D. Keilin, *The History of Cell Respiration and Cytochrome*, Cambridge, 1966.
9 J. Senebier, *Rapports de l'Air avec les Êtres Organisés ou Traités de l'Action du Poumon et de la Peau des Animaux sur l'Air Comme de Celle des Plantes sur ce Fluide. Tirés des Journaux d'Observations et Expériences de Lazare Spallanzani, avec Quelques Mémoires de l'Editeur sur ces Matières*, 3 Vols., Geneva, 1807.
10 W. F. Edwards, *De l'Influence des Agents Physiques sur la Vie*, Paris, 1824 (English translation in 1832).
11 G. von Liebig, *Arch. Anat. Physiol.*, (1850) 393.
12 M. Traube, *Arch. Pathol. Anat. Physiol.*, 21 (1861) 386.
13 C. Bernard, *Leçons sur les Propriétés Physiologiques et les Altérations Pathologiques des Liquides de l'Organisme*, Paris, 1859.
14 F. Hoppe, *Arch. Pathol. Anat. Physiol.*, 23 (1862) 446.
15 G. G. Stokes, *Proc. Roy. Soc.*, 13 (1864) 355.
16 P. Bert, *Leçons sur la Physiologie Comparée de la Respiration*, Paris, 1870.
17 C. F. Schönbein, *Ann. Physik*, 50 (1840) 616.
18 C. F. Schönbein, *Ann. Physik*, 67 (1845) 99.
19 C. F. Schönbein, *Verhandl. Naturforsch. Ges. (Basel)*, 1 (1856) 339.
20 C. F. Schönbein, *J. Prakt. Chem.*, 67 (1856) 496.
21 A. Schmidt, *Über Ozon im Blute. Eine physiologisch chemische Studie*, Dorpat, 1862.
22 O. Hammarsten, *Ber. Verhandl. Kgl. sächs. Ges. Wiss.*, 23 (1871) 617.
23 E. Pflüger, *Arch. Ges. Physiol.*, 6 (1872) 43.
24 E. Pflüger, *Arch. Ges. Physiol.*, 10 (1875) 251, 641.
25 S. Wolffberg, *Arch. Ges. Physiol.*, 6 (1872) 23.
26 G. Strassburg, *Arch. Ges. Physiol.*, 6 (1872) 65.
27 D. Finkler, *Arch. Ges. Physiol.*, 10 (1875) 368.
28 E. Oertmann, *Arch. Ges. Physiol.*, 15 (1877) 381.
29 A. Schmidt, *Arch. Pathol. Anat. Physiol.*, 42 (1868) 249.
30 H. von Helmholtz, *Arch. Physiol.*, (1848) 144.
31 C. Bernard, *Leçons sur la Chaleur Animale, sur les Effets de la Chaleur et sur la Fièvre*, Paris, 1876.
32 F. Hoppe-Seyler, *Z. Physiol. Chem.*, 14 (1890) 372.
33 C. A. Culotta, *Bull. Hist. Med.*, 44 (1970) 109.

Chapter 9

The Reaction Against "Analysm": Antichemicalists and Physiological Chemists of the 19th Century

1. Antichemicalism

During the first half of the 19th century, a number of chemists became known as organic chemists. They developed the consequences of the "chemical revolution", recognizing respiration as a combustion process and applying to the organisms the methods of elementary analysis introduced by Lavoisier and developed by several authors, among them Liebig. The establishment of the empirical formulas of the proximate principles isolated from organisms exerted a great appeal at the time when the four elements had been definitely confined to oblivion and replaced by the chemical elements. The organic chemists prepared many proximate principles and accumulated many data relating to the elementary analysis not only of proximate principles, when it was possible to isolate them, but also of bulk material obtained from organisms. From such data they derived metabolic theories, exemplified by the views of Liebig.

But this product of the chemical revolution was far from being considered as a happy one by many scientists at the time, as shown by the use of polemic and derogatory expressions such as "chemicalism" used, among others, by Caldwell, or "chimisme" coined by Bernard. This antichemicalism is not a self-consistent doctrine. It covers different forms of opposition to the body of theories built by the "analysts" and to their methodology. The ever-recurring conflict between vitalists and mechanists had not yet died out, and one aspect of antichemicalism, the vitalist one, opposed the chemists in the form of the antagonism towards any chemical theory of life which happened to be formulated at the time. Some chemists also believed that chemistry could not provide an explanation of life, on account of the latter's complexity.

References p. 189

Plate 34. Charles Caldwell.

The opposition of the physiologists was different. It consisted in a difference of opinion regarding methodology, based on the "self-consistent and all-embracing theory of the organism so beloved to biophysicists in the 1840s and 1850s"[1] and on the conviction also based on the naive mechanicist deduction that vivisection was the royal way to unravelling the complexities of the animal organism, including the chemical complexities. Physiological chemistry was the outcome of this form of opposition.

2. Opposition from the vitalist camp

Klickstein[2,3] has emphasized that, at the beginning of the 19th century, a movement had arisen at the University of Pennsylvania, which cast doubt on the importance of chemistry to medicine. This importance had been stressed by Benjamin Rush, one of the first advocates in Philadelphia of the utility of chemistry to medicine. Among those who adopted the opposite view was Charles Caldwell (1772–1835), a professor of natural history, and in spite of his insignificance as a scientist, it is useful to recall his views in detail as an example among many, of a common mentality of the time. In a lecture delivered in Philadelphia at the Medical School, in November 1818, Caldwell proclaims:

"Shall I be told, that chemistry aids, in the explication of any of the phenomena or laws of living body, either in a healthy or a diseased state? That it sheds light on physiology, pathology, or therapeutics? From the most correct and satisfactory views I have been able to form on this subject, I feel myself compelled to deny the position. As far as chemistry has mingled in discussions of this nature, it has not only darkened them, but filled them with error. It has superadded corruption to what it found already sufficiently corrupted". (ref. 4)

Caldwell left Philadelphia in the fall of 1819 and took the chair of medicine and clinical practice at Transylvania University, Lexington, Kentucky. In 1823, Caldwell published an outline of his lectures[5] of which a citation gives the tone. It is taken from the section *Chemical Theory of Life*:

"Chemistry, a science most signally delusive.
Instances of this, in the search after the philosophers' stone, the elixir vitae, the panacea, etc.
No less so in its application to physiology.
Instances—*Thought* declared by a European chemist to be the result of a chemical process.
By a writer, in our country, thc practicability of "*chrystallizing a man*" by a certain chemical process, openly asserted.
Anecdote of two lovers, one a chemist, the other a barrister, with its result.
Others embrace a half-way doctrine, and allege that *certain functions* only are of a chemical character, *viz.* digestion, respiration, secretion, absorption.
Objection. If life be chemical, man is degraded to the level of a laboratory.

References p. 189

Theory analyzed.
Chemistry defined.
Is a *circumscribed*, because it is purely an *experimental* science.
Had effected certain things, and nothing more.
Chemists not authorized in asserting a power to achieve beyond what they actually have achieved—Must not thus speculate *in advance*, and call their speculations science—They are *loose conjectures*, and nothing more—Because they have made glass, must not therefore, declare that they can make a diamond—nor assume the power to make living vegetables or animals, because they have constructed the *arbor Dianae*⋆.
Because, by chemistry *many* changes are effected in the decomposition and recomposition of matter, must not, therefore, declare that *all* are
This is to limit the powers and means of nature, and even to allege that she produces a plurality of effects, by a unity of cause—a doctrine wholly untenable.
Chemistry cannot produce results, without the presence and use of the requisite materials, in a *formal state*." (ref. 5)

In the following section, he examines chemical physiology:

"Chemistry is not only unable to make an animal, or a vegetable, but cannot perform even the simplest animal function.
Digestion.
Chemistry cannot out of an alimentary mass, form chyme.
Out of chyme, cannot form chyle.
Out of chyle, cannot form blood.
Hence, totally incapable of what may be termed, the vital liquefying process.
Of the solidifying, equally incapable.
Out of blood, cannot form either muscle, fat, skin, hair, feathers, or bone—from a homogeneous *fluid* cannot produce a variety of heterogeneous solids.
The same blood which forms *bone*, forms also brain—the complicated and exquisite organ of the mind—chemistry cannot direct this.
Doctrine of *chemical secretion* equally unfounded.
Most secreted fluids—urine—gastric liquor—saliva—matter of perspiration—pancreatic juice—semen masculinum—comes from the arterial blood.
Of these, chemistry cannot form even the simplest.
They are the result of *vital organic*, not of *chemical* action.
From a homogeneous fluid, the blood, chemistry cannot form so many other fluids, essentially different from each other.
Absorption, in part also a liquefying process, cannot be performed by chemical means—Does not go on by capillary attraction.
Bone, muscle, fat, skin, etc., cannot, by chemistry, be converted into lymph.
Absorbents said to make this mutation by means of a certain solvent.
Objections. Altogether hypothetical—no such solvent ever discovered." (ref. 5)

The chapter on the *Chemical Theory of Life* ends as follows:

"No sooner has life forsaken organized matter, than chemistry invades it—not before.
Chemical action destroys muscular strength and substance—*Vital* action preserves and augments them.
Vital action builds up organic matter; *chemical* pulls it down." (ref. 5)

⋆ A precipitate of silver amalgam, in the form of arborisations (author's note).

The doctrine of Caldwell again developed in 1829[6]. He defines the three schools, the chemical, the vital and the chemico-physiologic or chemo-vitalist, (represented for instance by Liebig) which relate to the changes occurring in living matter:

"The first alleges that, both in its origin, changes and dissolution, living matter is completely under the influence and governance of chemical principles; and that, therefore, in reality, every living body is nothing more than a chemical laboratory.

"The second hypothesis places living matter, with all the changes and phenomena it exhibits, under the exclusive control of principles denominated vital. Those principles and laws it pronounces to be insulated, and, in all respects, as essentially different, and as radically distinct from the principles and laws of chemistry, as they are from those denominated mechanical.

"The third hypothesis professes to be a compromise between the other two, and like most midway courses, is time-serving and unsteady. Its advocates are aptly denominated *chemico-physiologists*. This term is appropriate, and denotes very happily the character of those to whom it is applied." (ref. 6)

Caldwell is clearly a follower of the organo-vitalist school, and for the "chemico-physiologists" he predicts that "their hypothesis will be dissipated, by the increasing lights of science, as the shadows of night retreat from the sun."[6]

Caldwell was dismissed from the Transylvania Faculty in 1837. He went to Louisville where he took part in the foundation of Louisville Medical Institute, which became in 1845 the Medical Department of the University of Louisville. When Liebig's *Animal Chemistry* appeared in 1842 and was highly praised, it found its chief American opponent in Caldwell who published a pamphlet[7] against it in 1843. This was a quarrel between vitalists: Caldwell, an organo-vitalist, and Liebig, a chemico-vitalist. His prefatory remarks end with a short dialogue between a chemicalist (a word he coins), and a vitalist. The vitalist asks the chemicalist to answer a question:

"Do I understand the gentleman who has just concluded his discourse, to contend, that every function and every functional product of living organized matter are the results of *chemical agency*? Such, Sir, is my doctrine, replied the chemicalist, with a tone of confidence, and an air and look in no ordinary degree consequential and sapient. And then came on the pith and point of the contest.

"The vitalist, lifting a small covered dish, removed the lid, presenting it to his opponent, said: here, Sir, is a small quantity of well-prepared beef, potatoes and gravy—Will you oblige me by taking it into your laboratory, and changing it into genuine chyme? With an air of embarrassment, the chemicalist replied—really, Sir, I cannot, otherwise by eating it. That I can do myself, returned the challenger and a *pig* can do the same, as well as either of us." (ref. 7)

The pamphlet concludes with the pledge of Caldwell to fight the "unhallowed usurpation and contamination influence of dislocated chemistry".

The publication of Caldwell's pamphlet was followed by a vivid polemic

References p. 189

which is ably epitomized by Klickstein[4] from whose extensive study on Caldwell the citations of his works have been taken.

Even if Caldwell had supporters, the majority of the reviews of his pamphlet were derogatory, another symptom of the popularity of Liebig. None of his champions dared to insinuate that *Animal Chemistry* was a mere collection of errors. It is true that Caldwell's offensive was first sent from the decaying fortress of vitalism. In his answer to one of the derogative reviews, written by Yandell[8], Caldwell[9] takes an entirely different position. He quotes those who contradicted Liebig on factual evidence: Dumas and Boussingault in France, Pereira in Great Britain, Graves and Strokes in Ireland, and Hare in Philadelphia.

"In his introductory lecture of last autumn (1843) that distinguished gentleman proclaimed that in certain instances Liebig's writings were "*bold, hasty, inconsiderate*, and *inaccurate*" and that "the philosopher of Giessen has deluded himself and the readers of his essays with the prospect of an elucidation of the mysteries of animal and vegetable physiology, *which is beyond the present state of chemistry to afford.*" (ref. 9)

For an organo-vitalist like Caldwell, life cannot be explained by chemistry. Chemistry can only deal with dead matter: that which is left after life has gone. This was also the view of Barthez and of Bichat.

A high priest of the positivist cult, Emile Littré, of great dictionary fame, in a long review of a book published by Robin and Verdeil[10], concludes as follows, about the relation of physiology and chemistry:

"As everybody knows, there is an organic chemistry, *i.e.* a chemistry which deals with the organized substances. This expression must be clearly defined. If we mean by it that the organic phenomena, in the aspect of their simultaneous composition and decomposition, belong to chemistry; that the substances actually submitted to this act of composition and decomposition are chemical substances; that the action by which these substances maintain themselves in spite of the continuous combination and decombination are chemical actions, we fall in error and accept a false view of chemistry as well as of biology. There is no organic chemistry in this sense; there are superior properties, a superior molecular constitution which, in spite of depending, for its existence on chemical actions, is by no means their consequence, that is to say that we would vainly imagine any extension of the chemical phenomena; to whatever ideal limit they may be carried, they would never change themselves into vital phenomena. If, on the contrary, we mean that, once retired from the bodies and deprived of life, that is to say no longer presenting any molecular flux, the organic substances, from plants or animals, present nothing which prevents including them in the field of chemistry, we are right, and in this sense, there is an organic chemistry, full of difficulty and interest. It is death which transfers these substances from one domain to the other; but life, as long as it has manifested its breath, (*souffle*) has created, precisely because it belongs to a superior order, combinations of a higher complication as well and goes beyond, in this aspect, everything which can be seen elsewhere. Life has therefore elaborated in advance a field prepared for chemistry, a field which forces it to draw on itself and try all kinds of ways to lead its theories through this labyrinth. In this way the dividing line between chemistry and biology is accomplished: the dead organic substance belongs to the first; the living organic substances belongs to the second." (ref. 11, translation by author)

These lines clearly reflect the nominalist and antireductionist views of the positivist doctrine of Auguste Comte, expressed in his *Cours de Philosophie Positive*[12] and which commands the rejection of the cell theory as well as the theories of the "analysts". It is clear that the limitation of chemistry to the domain of death is not at all based on the same concepts as those held by Caldwell. Comte considered the theories of the "analysts" as well as the cell theory as going beyond the coordination of experimental facts, and he paradoxically landed on the same ground of such extreme vitalism as that professed by Caldwell.

3. Opposition from the chemists

Chaptal, professor of chemistry in Montpellier, and responsible for the denomination of nitrogen, was a convinced follower of Lavoisier but nevertheless gives an example of those chemists who thought that life escapes the laws of chemistry and believed that the variable and plastic processes taking place in organisms could not be accounted for by the complete determinism reigning in the rigid processes of chemistry and physics. Chaptal writes, at the end of the 18th century:

"But the more the functions of the individual are concerned with its organization, the less is the empire of chemistry over them and it becomes us to be cautious in the application of this science to all phenomena which depend essentially on the principles of life." (Chaptal[13], translation by Nicholson[14])

4. Opposition from the physiologists

This opposition originates in a critical examination of the methods and concepts of which the chemists made use in their endeavour to unravel the metabolic changes by the consideration of the elementary composition of the constituents of organisms. This episode, referred to above as "analysm", appears as the first source of the controversies between the chemists who believe they can solve the problems of biochemistry by a mere reference to chemical experimentation *in vitro*, and those who have a first-hand knowledge of organisms. This controversy still obtains among us. The present writer, at that time President of the newly founded International

References p. 189

Union of Biochemistry★ went to Oslo in 1955 to present the case of the application of that Union for membership in the International Council of Scientific Unions (ICSU). The theme of his presentation was the avocation of biochemistry as an experimental autonomous discipline with its own problems and concepts, and endeavouring to unravel, by adequate methods, the complicated pathways of cellular catabolic and anabolic metabolism. One of the opponents in the Council, a chemist, retorted that he could not understand why anybody should put in doubt his biochemical competence, as he himself performed the chemical synthesis of substances also biosynthesized by plants.

According to Northrop[15], there is a common pattern in all controversies on biochemistry.

"There is a complicated hypothesis, which usually entails an element of mystery and several unnecessary assumptions. This is opposed by a more simple explanation, which contains no unnecessary assumptions. The complicated one is always the popular one at first, but the simpler one, as a rule, eventually is found to be correct. This process frequently requires 10 to 20 years. The reason for this long time lag was explained by Max Planck. He remarked that "scientists never change their minds, but eventually die." Of the constancy of controversies between biologists and chemists, Northrop gives the following explanation: "The biologist uses the language of semantics and is satisfied with a descriptive and qualitative theory. The chemist leans towards the language of mathematics and requires quantitative results. There is still another difference: chemists (and physicists) have a great respect for the Reverend Occam's razor★★ and endeavour to limit their assumptions to the minimal number essential for an explanation, in accordance with the principle of the conservation of hypotheses; whereas some biologists have no respect for the Reverend's weapon and fearlessly bolster an ailing (and unnecessary) assumption by another similar one. As a result, the chemist, who thinks he stands on firm ground, is frequently astonished to find himself facing a whole company of unnecessary assumptions, which he is expected to disprove, rather than lop off with William of Occam's weapon". (Northrop[15])

It is clear that the progress of scientific biochemistry was for a great part based on chemical knowledge and on chemical paradigms, but one of the scopes of this History is to reach an identification of the methodological directions through which biochemistry, by the use of several approaches to the knowledge of molecular constituents of the organisms, has developed the conceptual system of the interpretation of its field of study, which has by the use of a number of different methods, besides the chemical one, extended progressively to all the aspects of the nature of living creatures and has become the universal basis of a science of life. In this context the

★ See Appendix 1 (p. 321).

★★ A method of argumentation adopted by William of Occam, consisting of eliminating all unnecessary facts or constituents for a question under analysis.

biochemists are those who have taken advantage of Occam's razor. The chemist followers of Lavoisier—Fourcroy, Hallé, Prout and above all Liebig—made an excessive use of this razor, and their endeavour to unravel pathways by "writing-table biochemistry" based only on the empirical formulas of the chemical compounds found in organisms, went on the rocks.

Immediately after the publication of Liebig's *Animal Chemistry*, his conclusions and his methods were criticized in the physiologists' camp. The treatise of Kohlrausch[16] showed that many of Liebig's views were unsubstantiated, such as the concept of the derivation of the whole of animal heat from the combustion of carbon and hydrogen, that the bile is reabsorbed and consumed in respiration or that the non-nitrogenous components of food are never incorporated into organized tissues. Kohlrausch pointed out that, using Liebig's methods, one could start from any formula containing carbon, hydrogen, oxygen and nitrogen and arrive at any other formula by adding or subtracting the elements of ammonia, water and oxygen. By calculations analogous to those performed by Liebig he showed that one could at will reach the conclusion that the oxidation of protein in the body produces either the poison prussic acid or fulminic acid, an explosive compound.

In 1853, Robin and Verdeil[10] published a copious work of three volumes in which the method of the biologists was opposed to the method of the chemists of the time. The authors rightly cite as example the fact that for the chemist, urea was the oxidation product of uric acid, while for the physiologist, it was the endproduct of amino-acid nitrogen metabolism. The "writing table biochemistry" based on elementary composition is here directly attacked. As the authors say: the chemists' way of reasoning amounts to saying that if we add oxygen to the formula of fibrin we obtain the formula of uric acid.

According to the authors, the "proximate principles" are not, as Chevreul thought, defined chemical species, not being considered as crystallizable. Therefore their peculiarities and modifications should be followed and studied by the biologist as well as muscle fibres or nervous tubes, for instance. The "proximate principles", as well as "cells and fibres", were considered as parts of "organized matter", in the sense of Bichat.

Bernard's antichemicalism was of a different nature. He objected to theories built without recourse to experiment on animals and only on the basis of the comparison of the elementary composition, for example, of a substance introduced into the organism and an excretion product. What he

References p. 189

fought was not chemistry but an unjustified speculation derived only from elementary composition as indulged in by Fourcroy, Hallé, Prout and Liebig. In fact he was one of the main chemically-inclined physiologists of his time. Like Kohlrausch, Bernard claims the necessity for experimenting on organisms in order to know the chemical processes of which they are the seat. In his last book, *Leçons sur les Phénomènes de la Vie*, Claude Bernard[17] clearly explains his views:

"The chemistry of the laboratory and the chemistry of the living body obey the same laws; there are not two chemistries, Lavoisier said this. But the chemistry of the laboratory is accomplished by using agents and apparatus created by the chemist, and the chemistry of the organism is accomplished with the help of agents and apparatus created by the organism. The chemist, for instance, transforms starch into sugar with the help of an acid he has prepared; he saponifies fats with the help of soda lime, of concentrated sulphuric acid, of superheated water vapour, all agents he has prepared himself. The animal, as well as the germinating seed, transforms starch into sugar without using an acid, with the help of a ferment (diastase) which is a product of the organism. Fat is saponified in the animal organism, in the intestine, without the participation of any soda lime, without superheated water vapour, but with the help of pancreatic juice which is a secretion produced by a gland." (ref. 17, translation by author)

In the preparatory notes, published by Delhoume for his *Principes de Médecine Expérimentale*, Bernard[18] writes:

"Whatever the idea we may have of the role of vital phenomena taking place in living organisms, we must consider them as physico-chemical phenomena of a special nature which cannot be studied except in the very organism, so to speak." (ref. 18, translation by author)

When he refers to vital phenomena or vital properties Bernard is not indulging in any philosophical theory and he simply expresses reservations towards naive chemical reduction. As he writes in the *Leçons sur les Phénomènes de la Vie*[17], "The vital properties are in reality nothing but the physico-chemical properties of organized matter" and the organized matter is itself defined as follows.

"The organization results from a mixture of complex substances, reacting one on the other. For us it is this arrangement which gives rise to the immanent properties of living matter, this arrangement being special and very complex, but obeying nevertheless the general chemical laws of the aggregates of matter." (ref. 17, translation by author)

Voit has also clearly stated the principle of experimentation on animals as the basis of biochemistry in his criticism of Liebig's *Animal Chemistry*:

"But he has forgotten, to the sorrow of those who know and value his high services to science better than do his flatterers, that these are all mere ideas and possibilities, whose validity must first be tested through investigations on animals; and this is the basis on which I have placed myself." (ref. 19, translation by Holmes)

The same idea was already formulated by the chemist and naturalist C. F. Schönbein, who in 1863 remarked that in very few cases only

> "vom Chemismus des Laboratoriums auf denjenigen der lebendigen Natur geschlossen werden konnte und man leider von dem Erfolg unserer mühevollsten Arbeiten dieser Art mit dem Dichter nur zu oft sagen muss: "Zum Teufel ist der Spiritus, das Phlegma ist geblieben". Es muss deshalb aüsserst wünschenswert erscheinen, Mittel und Wege der Forschung auszufinden, mehr als die bisherigen geeignet, uns zum Verständnis der so feinen chemischen Vorgänge zu führen, welche in der lebendigen Tierwelt und Pflanzenwelt stattfinden". (ref. 20)

5. The era of physiological chemistry (*ca*. 1840–*ca*. 1880)

The organic chemists, besides isolating proximate principles, characterizing them and determining their empirical formulas, overvalorized their approach in such a way as to lead to an enormous accumulation of analytical data concerning even the elementary analysis of whole organs, data that were supposed to lead to the elucidation of the nature of diseases. When this approach was recognized as inadequate, organic chemistry followed the path that will be retraced in Chapter 12 and led to the development of a study of carbon compounds and of the modern science of organic chemistry along the paths of organic synthesis and of the expression in structural formulas, of the properties and of the structure and shape of the natural organic molecules.

The toils of the "analysts" with respect to organisms having led to confusion, the physiologists, who had been active in denouncing this failure, annexed the chemical study of organisms cast off by the chemists, as a section of physiology, under the denomination of Physiological Chemistry. Such study is sometimes said to be a combination of physiology and of chemistry, but this definition falls short of giving the proper specific character of the new discipline, which aimed to solve biochemical problems by recourse to the method of vivisection. It is in this sense that we will use the denomination which is sometimes used in a loose way nowadays to mean the molecular basis of physiological functions. At this time, *i.e.* in the 1840s and 1850s, the physiologists were dominated by a naive mechanicist reductionism, a sequel of iatromechanics. Liebig is sometimes considered as one of the physiological chemists, and even sometimes as the first one; whereas, as shown in the previous chapters, he was the most extreme of all "analysts" and never performed a physiological experiment. Perhaps the statement derives from the title of his book: *Animal Chemistry, or Organic*

References p. 189

Chemistry in its Applications to Physiology and Pathology[21], and from the wish he formulates in his introduction to that book:

"Physiology took no share in the advancement of chemistry, because for a long period she received from the latter science no assistance in her own development. This state of matters has been entirely changed within five and twenty years. But during this period physiology has also acquired new ways and methods of investigations within her own province; and it is only the exhaustion of these sources of discovery, which has enabled us to look forward to a change in the direction of the labors of physiologists. The time for such a change is now at hand; and a perseverance in the methods lately followed in physiology would now, for the want, which must soon be felt, of fresh points of departure for researches, render physiology more extensive, but neither more profound nor more solid . . .

. . . In the hands of the physiologist, organic chemistry must become an intellectual instrument, by means of which he will be enabled to trace the causes of phenomena invisible to the bodily sight . . ." (translation of Gregory[21])

The physiological chemists knew chemistry after a fashion, and their experimental approaches at the level of the whole complexity of organisms were almost limited to studies on dogs and rabbits. Dominated, as they were, by the overvalorized individuality of the different organs to which they attributed specialized functions, they overrated these aspects and failed to recognize the existence of a central biochemical system common to all cells, and consequently were not able to define the molecular basis of the activities of differentiated cells. Physiological chemistry has nevertheless added to biochemical knowledge such important contributors as the indirect nutrition concept and the notion of the internal environment, both introduced by Bernard, and it has ruled out the ever and anon recurring "direct assimilation", as will appear in Chapter 10.

One of the most influential treatises of physiological chemistry was published by Lehmann in three volumes between 1849 and 1852, and a second edition was already published in 1853[22]. Lehmann's introduction may be taken as formulating the conceptual basis of the new science. He criticizes the conclusions based on the simple consideration of empirical formulas.

"Darum sind auch in der Physiologie, hauptsächlig aber in der Pathologie, und der Stelle der früheren naturphilosophischen Phantasien eine Menge chemischer Fictionen getreten, durch welche die Medizin in ein neues Labyrinth der haltlosesten Theorien gestürzt worden ist." (ref. 22)

He sharply criticizes the conclusions based on elementary analysis of mixtures of substances, a method widely used by pathologists. He points to the imperfect state of the techniques of analysis, which renders the data unreliable, and to the necessity of improving these techniques. He insists on the point that the physiological chemists must amalgamate the chemical and the physiological data,

"d.h. die einzelnen Stoffe müssen nicht nur rücksichtlich ihres vollen chemischen Werthes und nach ihrer Stellung in der reinen organischen Chemie in allen Richtungen genau erforscht werden, sondern ihre Behandlung muss zugleich auch bereits die allgemeinern Beziehungen in sich aufnehmen, welche jeder einzelne Stoff für den thierischen Organismus und dessen Stoffwechsel haben kann. Mit einem Worte, in der Zoochemie, als Grundlage der physiologischen Chemie, ist der physiologische Werth jedes einzelnen Stoffes ebenso sorgfältig darzulegen, als seine Stellung und Geltung im Systeme der reinen Chemie." (ref. 22)

In the first volume (Zoochemie) of the treatise the organic compounds of the animal body are enumerated, and for each substance the properties and preparation are followed by its physiological value (*Werth*) and what is known or supposed about its metabolism. For instance, when he considers urea he presents it as deriving in the animal body from uric acid. This study of the constituents (zoochemistry) leads to a proper understanding of tissues and fluids (*Saftenlehre*) of the organism and to the zoochemical processes, *i.e.* the science of metabolism. The methodology of this study is discussed, and Lehmann records three methods. The first consists of establishing a balance sheet between what enters the organism and what leaves it. A second is to reproduce *in vitro* the phenomena taking place in organisms and draw analogies between both sets of observations. The third consists of true physiological experiments on the whole organism.

The first method—comparing ingesta and excreta—was applied by physiological chemists, for instance, when it was shown that in an adult dog the amount of nitrogen excreted corresponds to the amount of nitrogen ingested. It may be stressed that the method is of no use if the compound introduced in the food is completely catabolized to CO_2 and H_2O. On the other hand the experimental method on an animal may be perfected by a certain recourse to surgery. For instance, while a normal mammal transforms the amino acid nitrogen into urea nitrogen this no longer occurs if its liver has been removed. Another development of the approach consists of the use of isolated organs. If ammonia is introduced into the perfusion fluid of an isolated liver, it leads to the presence of urea in the efferent fluid. We shall have opportunities later in this History to consider a number of these direct organismic approaches (some of them of great importance) when we deal with the study of the history of the molecular bases of the physiological or biological concepts. A large collection of disconnected data resulting from the analysis of animal tissues and body fluids, from the study of

References p. 189

Plate 35. Felix Hoppe-Seyler.

chemical models, or from experimental approaches on the whole organism or some of its parts, was made by the physiological chemists.

6. Journals devoted to physiological chemistry

In 1872 Hoppe-Seyler was appointed to the first chair of physiological chemistry, at Strasburg University. In 1877, he created the first periodical in the field, under the name of *Zeitschrift für physiologische Chemie*, devoted to the study of the "chemische Lebensvorgänge"[23]. The members of the editorial board were Baumann (Leipzig), Gähtgens (Rostock), Gorup-Besanez (Erlangen), Hüfner (Tübingen), Huppert (Prague), Maly (Graz) and Salkowski (Berlin). This was little appreciated by the majority of physiologists who were used to considering physiological chemistry as a mere handmaiden of physiology. Pflüger[24] raised a protest and claimed that his *Archiv* would continue publishing papers on physiological chemistry. Hoppe-Seyler[25] answered this by claiming the right of physiological chemistry to have a place of its own in research and teaching, and to prepare the students to consider the proper importance, in their curriculum, of physiological chemistry which was previously neglected owing to those physiologists who either ignored chemistry or failed to understand its importance. Pflüger had warned against tendencies to split physiology. To this Hoppe-Seyler retorted that the real danger for a scientific discipline is the despotic abasement of forces which could, when free, lead to new developments. The sting in the tail was a transparent allusion to the low degree of consideration given by Pflüger's periodical to papers of a biochemical nature.

"Diese Zeitschrift soll nur die neuen physiologisch–chemischen Arbeiten sammeln und dem wissenschaftlichen Publikum vorführen, was in dieser Richtung neues geleistet wird; dass auch dies nicht ohne Hindernisse und Schwierigkeiten geschehen kann, hat mir um so weniger unbekannt bleiben können, als mir von befreundeter Seite mehrfach solche Hindernisse bezeichnet worden sind." (ref. 24)

7. From physiological chemistry to biochemistry

The weak point of physiological chemistry layed in the endeavour to solve biochemical problems *in vivo* by purely physiological methods before the fundamental properties of the molecules involved were determined. When organic chemistry took a different path, and when accurate knowledge was provided by its progress, a new approach, no less experimental, but taking

References p. 189

into account the nature, the structure and the properties of the molecules concerned, developed progressively, a process in the course of which biochemistry accomplished its separation from physiology.

One of the pioneers of this tendency was Hoppe-Seyler, and his move when founding the *Zeitschrift für physiologische Chemie* in 1877 was the first act of independence of what, with time, was to become biochemistry. The *Zeitschrift für physiologische Chemie* and the *Beiträge zur chemischen Physiologie und Pathologie* were the only journals whose titles mentioned physiological chemistry. When enzymology, during the first decade of the 20th century, found its first successes in the identification of the intermediate steps of a microbiological phenomenon, alcoholic fermentation, the interests of chemically-inclined biologists gained breadth and extension and the denominations of "Biological Chemistry" and "Biochemistry" became more generally adopted, as "Physiological Chemistry" in a large degree limiting the field of enquiries to animal chemistry while plant chemistry and microbiological chemistry were recognized as revealing more fertile fields for research, which could be approached by a variety of methodologies.

The *Journal of Biological Chemistry* was started in 1905 and so was the *Biochemical Journal.* The *Biochemische Zeitschrift* appeared in 1906, and the first issues of the *Bulletin de la Société de Chimie Biologique* appeared in 1914, the publication being interrupted by the 1st world war but resumed in 1918.

The controversies retraced in this chapter recall the historical background of the occasional persistence of opposition between biochemists on the one hand, and chemists or physiologists on the other hand.

Chemists have had a tendency to consider biochemistry as a study of slime, and physiologists once used to consider physiological chemistry as a mere servant of physiology. These oppositions have disappeared, more or less completely, with time, owing to the factor pointed out by Planck and mentioned above: if scientists never change their minds, they eventually die.

REFERENCES

1 R. Olby, XIIe Congrès International d'Histoire des Sciences, Paris, 1968, *Actes*, 8 (1971) 135.
2 H. S. Klickstein, *Library Chronicle*, 16 (1950) 64.
3 H. S. Klickstein, *Bull. Hist. Med.*, 27 (1953) 43.
4 H. S. Klickstein, *Chymia*, 4 (1953) 129.
5 Ch. Caldwell, *Outlines of a Course of Lectures on the Institutes of Medicine*, Lexington, Ky., 1823.
6 Ch. Caldwell, *Transylvania J. Med. Assoc. Sci.*, 2 (1829) 1.
7 Ch. Caldwell, *Physiology Vindicated, in a Critique on Liebig's Animal Chemistry*, Jeffersonville, Ia. (now Indiana), 1843.
8 L. P. Yandell, *Western J. Med. Surg.*, 8 (1843) 430.
9 Ch. Caldwell, *Western J. Med. Surg.*, (2) 1 (1844) 1.
10 Ch. Robin and F. Verdeil, *Traité de Chimie Anatomique et Physiologique Normale et Pathologique*, 3 Vols., Paris, 1853.
11 E. Littré, *Revue des Deux-Mondes*, 1 jan. 1855. [This article entitled "De la science de la vie dans ses rapports avec la chimie" has been reprinted in E. Littré, *La Science au Point de Vue Philosophique*, 3rd ed., Paris, 1873.]
12 A. Comte, *Cours de Philosophie Positive*, 6 Vols., Paris, 1830–1842.
13 J. A. Chaptal, *Éléments de Chimie*, 3 Vols., Montpellier, 1790.
14 J. A. Chaptal, *Elements of Chemistry*, 3 Vols., translation by W. Nicholson, London, 1791.
15 J. H. Northrop, *Ann. Rev. Biochem.*, 30 (1961) 1.
16 O. Kohlrausch, *Physiologie und Chemie in ihrer gegenseitigen Stellung, beleuchtet durch eine Kritik von Liebig's Thierchemie*, Göttingen, 1844.
17 C. Bernard, *Leçons sur les Phénomènes de la Vie Communs aux Animaux et aux Végétaux*, Vol. 1, Paris, 1878.
18 C. Bernard, *Principes de Médecine Expérimentale (Introduction et Notes par le Dr. Léon Delhoume)*, Paris, 1947.
19 E. Voit, *Z. Biol.*, 6 (1870) 399.
20 C. F. Schönbein, *Sitzber. Kgl. bayer. Akad. Wiss.*, 2 (1863) 118.
21 J. von Liebig, *Animal Chemistry* (translation by W. Gregory), Cambridge, 1842.
22 C. G. Lehmann, *Lehrbuch der physiologischen Chemie*, 3 Vols., Leipzig, 1849–1852 (2nd ed., 1853).
23 F. Hoppe-Seyler, *Z. Physiol. Chem.*, 1 (1877) Vorwort.
24 E. F. W. Pflüger, *Arch. Ges. Physiol.*, 15 (1877) 25.
25 F. Hoppe-Seyler, *Z. Physiol. Chem.*, 1 (1877–1878) 270.

Chapter 10

Dynamic Permanence after Liebig, and the End of the Myth of Direct Assimilation

1. Composition and decomposition at the cellular level

Liebig's views, as we saw in Chapter 7, greatly simplified the concept of assimilation in animals. For him, proteins were the only nutrients to be assimilated by animals, the others being burned in blood to provide animal heat. Since, according to him, the proteins remained practically unchanged during digestion and directly replaced the effete material of tissues, there was no problem of assimilation. When, after Liebig's system collapsed, the problem rose again, the current concept remained the notion of "composition and decomposition", which, after Schwann's formulation of the "theory of the cells", was considered this time as taking place at the level of each cell, as stated in his epoch-making book.

Composition and decomposition at the level of "elementary parts" is discussed in Robin and Verdeil's book[1] published in 1853. They consider that some substances are formed, and others decomposed, while certain substances last as long as the organisms, continually renewing the substance of which they are formed, by the loss and replacement of their constitutive elements, as for example in "albumin". According to these authors, these processes are simultaneous and continuous in composition and in decomposition. The same concept appears in the book of Robin[2] published in 1873. He describes how proximate principles penetrate by endosmosis into the cells and progress, molecule by molecule, by "intussusception" into the mass of each cell. These proximate principles combine with the substance of the cell and form compounds similar to those of the cell. This he calls "assimilation", by which some principles become similar to those characteristic of the cell. For each cell, some principles penetrate, some leave the cells, while others appear to be permanent constituents.

References p. 212

Plate 36. Charles Robin.

2. Theories of protein metabolism

Between 1867 and 1905 a number of theories of protein metabolism were proposed (for references, see Cathcart[3]). In 1867, the theory of Liebig was revived in a modified form when Voit proposed that protein of the food penetrated into the blood and became "circulating protein". This circulating protein, for the most part, was directly catabolized at the cellular level while a small part was used to repair the cell proteins damaged by wear and tear. In his views, the "circulating" protein did not differ from the "tissue" protein.

Contrary to previous theories, the basis of Voit's theory was experimental. He observed, for instance, that in starvation about 1 percent of the tissue proteins is broken down per day, while if proteins are fed in amounts equal to 12 percent of the "body proteins", the breakdown is 15 times greater than in starvation. From this he concluded that the catabolism of tissue protein was insignificant compared with the catabolism of circulating proteins introduced with the food. The opposite view was adopted by Pflüger (1893). For him, the tissue proteins are readily catabolized, while food proteins are not. Pflüger believed that there was a fundamental difference in constitution between tissue proteins and food proteins. He thought that the greater lability of the tissue proteins was the result of the presence of cyanogen radicals in their molecules.

The basis of Pflüger's theory was experimental, consisting of experiments made in his laboratory by Schöndorff. Blood was taken from a starving dog and circulated through the liver and hind limbs of a well-fed dog: the concentration of urea in this blood was increased. If blood from well-fed or starved dogs were circulated through the liver and hind limbs of a starving dog, no increase of urea was observed.

Folin (1905) criticized the conclusions on the basis of a quantitative discussion of the data. On the other hand, Pflüger's theory does not fit in with the fact that a rapid rise of output of nitrogen in the urine is observed after a protein meal.

According to the theory of Speck (1903), two forms of protein exist. The part of the food protein that is not used in the formation of new tissue is broken down into a nitrogen-containing part, rapidly converted into urea, and a nitrogen-free part, readily converted into a fat or carbohydrate-like substance. The nitrogen-containing part is converted into a great variety of substances, ultimately leading to urea, which is excreted.

References p. 212

Folin (1905) performed a large number of analyses on urine of human subjects on standard diets rich in nitrogen or poor in nitrogen but all of them practically free from purine, creatine and creatinine. He concluded that, whatever the amount of protein in the food, large or small, the amounts of creatinine and of neutral sulphur remain constant, as well as uric acid and ethereal sulphates, although to a lesser extent. On the other hand, when the amount of protein ingested was modified, changes in the excretion of urea and inorganic sulphates were observed. He concluded that there were two types of protein metabolism, one, variable, he called exogenous protein metabolism or intermediate metabolism, and the other, constant, he called tissue or endogenous protein metabolism.

Folin's theory was generally accepted until it became known that creatinine is a metabolite of creatine and not of proteins, and that in man uric acid is a product of nucleic acid metabolism and not of protein metabolism. In the light of these new concepts the whole theory collapsed.

The myth of "direct assimilation" was maintained as long as it was believed that the proteins of the food passed into the blood as such, or with little change. As soon as it was recognized, as we shall see later, that proteins were degraded in the intestinal tract, into amino acids which were brought to the tissues by the blood, the notion of assimilation, according to which a protein was introduced as such or after superficial changes into cells, was replaced by the concept of biosynthesis. Biosynthesis and catabolism replaced composition and decomposition. The concept was nevertheless still present in the form of the relative dependence of the two aspects, and the belief in independent exogenous and endogenous types of metabolism, part of the food being used for energy liberation, part for repair, the organism being considered as a frequently-repaired combustion engine. As in the concept of Robin, there persisted the view according to which the structural elements of cells are primarily in a stable state, the molecules brought by the blood being catabolized by the action of specialized enzyme systems and providing energy, while other parts were used by the permanent machinery to repair its worn-out parts.

For a long time it was maintained that biosynthesis was limited to plants. It was generally believed that animal cells were deprived of the capacity of biosynthesis. The rupture with this concept is due to Bernard.

When, in 1841, Bernard was appointed as "préparateur" in Magendie's laboratory, he became an adept of the vivisection method praised by his master, who had come to the front in physiology, but at the same time he worked in the laboratory of his friend, the chemist Pelouze. This, too, may have been a consequence of the influence of Magendie who held the opinion that chemical methods were of great use in physiological investigations. As stated by Schiller[4], the influence of Magendie on Bernard acted in two ways: through vivisection and through chemistry. Schiller, who has written an interesting description of the relations of Bernard with the laboratory of Pelouze and with two chemists at work in this laboratory, Barreswil and Margueritte, notes that Bernard could not have received any kind of chemical initiation from Magendie, who entirely relied on chemists when needed and that Bernard learned to master chemical methods by an assiduous contact with the laboratory of Pelouze between 1840 and 1850. It must also be noted that the first papers published by Bernard concerned animal chemistry. Following on Magendie's work on nutrition, it was Bernard's ambition to combine vivisection and chemistry in order to penetrate into the still occult field of nutrition in the workings of the organism. It must be recalled here briefly, as we will deal with the history of digestion in more detail in another part of this History, that Eberle[5], in 1834, had formulated the theory according to which the gastric mucus is the agent of the digestion of egg white in the stomach. Schwann[6] took a different position when he concluded from his experiments that gastric digestion is a fermentation phenomenon. Schwann had shown that hydrochloric acid is not the agent of the hydrolysis, and he concentrated the active principle which he named pepsin and which he showed could act in very small amounts compared with the amount of protein transformed. He pointed to the analogy with alcoholic fermentation, in which very small amounts of yeast act on large amounts of sugar, and for this reason proposed to consider the gastric digestion of proteins as a fermentation. But, he remarks, in the gastric process no alcohol is formed as in alcoholic fermentation, nor any acetic acid as in acetic fermentation. Besides, yeast does not degrade coagulated egg white. He concludes that the phenomena of alcoholic fermentation (of which he had not yet recognized at the time the agent, yeast, as of cellular nature), of acetic fermentation and of protein digestion must receive the common denomination of fermentations because they are the result of the action of very small quantities of active agents.

Bernard and Barreswil[7], in 1844, while adopting the action of the gastric juice on coagulated egg white as the result of the action of Schwann's pepsin, erroneously identify the gastric acid mainly as lactic acid, but in subsequent reports Bernard was to adopt an approach he expresses when he

References p. 212

writes:

"It is known that gastric juice is acid; the nature of the acid or of the acids which confer this acidity to this secretion is only of secondary interest, of which we may take no notice, but the acid reaction of the gastric juice is the dominant and characteristic property of this organic fluid." (ref. 8, translation by author)

We recognize in this sentence the attitude of physiologists when dealing with biochemical problems, but at the same time it must be underlined that in the course of his studies on digestive juices, Bernard was interested in the fate of nutrients in the animal body and he was convinced that it was not the right approach simply to compare the elementary composition of what enters the body and the endpoint of its changes during the catabolic process.

3. The glycogenic function of the liver

In the field of carbohydrate metabolism it was accepted that animals were only able to catabolize the sugar provided by plants. Combustion was considered as taking place in the blood. Bernard showed that animals were able to synthesize glycogen (glycogenic function of the liver). The history of this discovery, a milestone in the history of biochemistry, has been retraced by Grmek[9] on the basis of unpublished manuscripts. As Claude Bernard states in the *Introduction*[10] (p. 163):

"In 1843[12], in one of my first pieces of work I undertook to study what becomes of different alimentary substances in nutrition. As I said before, I began with sugar, a definite substance that is easier than any other to recognize and follow in the bodily economy. With this in view, I injected solutions of cane sugar into the blood of animals and I noted that even when injected in weak doses, the sugar passed into the urine. I recognized later that, by changing or transforming sugar, the gastric juice made it capable of assimilation (*assimilable*), *i.e.* of destruction in the blood (*destructible dans le sang*)." (ref. 10, translation of Greene[11])

It is clear that by the adjective "assimilable" Bernard refers to the fact that the sugar can be used by the organism and makes no reference to assimilation proper in the classical sense. He certainly does not write "that assimilation is synonymous with destruction in the blood" as Greene's translation may seem to mean. On this subject we read the following phrases of Greene's translation of the *Introduction*:

"I wished to learn in what organ the nutritive sugar disappeared, and I conceived the hypothesis that sugar introduced into the blood through nutrition (*l'alimentation*) might be destroyed in the lungs or in the general capillaries. The theory, indeed which then prevailed and which was naturally my proper starting point, assumed that the sugar present in animals came exclusively from foods, and that it was destroyed in animal organisms by the phenomena of combustion, *i.e.* of respiration. Thus sugar had gained the name of *respiratory nutriment*. But I was immediately led to see that the theory about the origin of sugar in animals, which served me as a starting point was false. As a result of the experiments which I shall describe further on, I was not indeed led to find an organ for destroying sugar, but on the contrary, I discovered an organ for making it, and I found that all animal blood contains sugar even when they do not eat it." (ref. 10, translation of Greene[11])

The steps which led him to give up the current theory are, as stated by Wilson[13], by Young[14] and by Olmsted[15], more detailed in Bernard's thesis for his degree of "docteur ès sciences", published in 1853. Bernard wanted to know whether the sugar absorbed from the food was destroyed in passing through the liver, or, in passing through the lungs, etc. He fed a dog on carbohydrates for seven days and killed it during digestion. He found that the blood of the hepatic veins, at the point where they join the inferior vena cava, contains a large amount of glucose. Wishing to perform the counter-proof called for by his methodological principles, he fed a dog exclusively on meat and was surprised once more to find large amounts of sugar in the blood of the hepatic veins, although the intestines did not contain sugar. Before he reached the conclusion that he was on the wrong track by following current theories, Bernard, as the perusal of his manuscript notes showed to Grmek[9], performed a large number of experiments concerning the location of the destruction of carbohydrates introduced into the organism with the food, by intravenous injection, etc.

In 1845 he turned his interest towards diabetes and concluded that it was a disease of the lungs. This conclusion rested on four premises: (*1*) sugar cannot be synthesized in the animal body; (*2*) it is normally destroyed in the lungs; (*3*) the principal symptom of diabetes is the presence of undestroyed sugar in the urine; and (*4*) the nervous system controls the breakdown of sugar in the lungs[9].

Glycaemia was at the time considered as an accidental or pathological aspect, and Bernard was the first to demonstrate that glycaemia was normal and independent of alimentation. This discovery was related in manuscript notes compiled between 1846 and 1848. Until May 1848 he was unable to progress, but at that time, since he believed that sugar must be destroyed somewhere, and as the lungs were supposed to be involved, he submitted grape sugar, *in vitro*, to contact with fresh lung tissue. After 10–12 h the sugar had disappeared. Bernard concluded that a special ferment for the degradation of glucose is present in the lungs. But, faithful

References p. 212

Plate 37. Claude Bernard.

to his methodology, he decided to perform a counter-experiment, mixing grape sugar with tissues of liver and of other organs. As he obtained the same results as with lung tissue he became puzzled with respect to the special function of the lungs in metabolizing sugar. On the last days of May 1848 he performed the following experiment: one gram of grape sugar was injected into the jugular vein of a dog while at the same time blood was taken from the carotid artery. This blood contained a large amount of sugar. The conclusion was that, as the blood must pass through the lungs to circulate from the jugular vein to the carotid artery, glucose is not destroyed in the lungs. Since he observed that, in the presence of fibrin, sugar does not react well with the copper reagent devised by his friend Barreswil, he imagined a new theory of diabetes. Sugar was supposed to be destroyed in the blood, fibrin playing a role in this destruction, and diabetes was an inhibition of this destruction by a chemical disorder of fibrin synthesis. As he wished to put this new theory to the test of experimentation, Bernard found that in dogs fed with a rich carbohydrate diet, the blood from the hepatic veins and vena cava contained sugar, as well as the blood of both ventricles of the heart: therefore neither liver nor lung destroyed sugar. Wishing to perform a counter-experiment, he did the same experiment with a dog given a non-carbohydrate diet and killed by section of the bulb. To his great surprise, he found that the blood of the portal system contained enormous amounts of sugar, while small amounts were found in the heart, only traces in the arterial blood and none in the chyle. The whole experiment contradicted all expectations. Bernard soon realized that this was caused by an artificial reflux of hepatic blood due to a suction effect of the opening of the abdomen. It was at that time, after the experiment of August 1848, that Bernard understood that the current theories were false and that the liver produced sugar. The real demonstration came in September 1848, when Bernard performed experiments on dogs with ligatures of the blood vessels and determination of blood sugar in different sectors of the circulation. In the portal blood of a dog starved or fed on meat, he found no sugar, while he found sugar in the blood of the suprahepatic veins. It has been appropriately said that Bernard was lucky, as his reagent was not sensitive enough to detect sugar in fact present in the portal blood, and also because his dogs were in a state of marked hyperglycaemia due to his method of killing them by sectioning the medulla oblongata. Nevertheless, his experiments made it clear that the liver was able to liberate sugar.

References p. 212

4. Indirect nutrition

The origin of this glucose was stated by Bernard in his Lectures[17] of 1855 when he indicated that the liver makes glucose with nitrogenous substances, a conclusion which was based on experiments prosecuted by Lehmann who had confirmed on a horse the experiments of Bernard and who had observed that the portal blood, when passing through the liver, loses proteins which were consequently considered as the source of the sugar.

It may be said that, at this time, Bernard was proposing a theory that was to revive later under the name of gluconeogenesis, and he opposed this theory to the one proposed by his opponent Figuier who claimed that blood sugar came from an accumulation, in the liver, of sugar coming from the food, a view which was radically rejected by Bernard, while Figuier, on the other hand, radically rejected the possibility, for the liver, of performing a synthesis of sugar from other chemical sources. The next step in the trail of research followed by Bernard was the discovery, the isolation and the purification of glycogen. He made a series of comparative analyses of liver taken immediately after death and after different periods of time. Perfusing a freshly removed liver through the portal vein with cold water for forty minutes, he showed that after that time, no sugar remained in the fluid that had perfused the liver. But when, owing to lack of time, he once kept a liver until the next day, before doing the analyses, he observed that the fluid which swept out of it contained sugar. This did not happen if the liver had been submitted to the action of boiling water. Bernard[18] presented these results to the Academy of Sciences on September 24, 1855, stating that liver contains an insoluble compound (glycogenic matter) which liberates sugar as the result of an enzymatic process (diastasis). In a lecture[19] delivered on 18 March 1857 and in a communication[20] to the Academy on 23 March 1857, Bernard reported on the isolation and purification of glycogen, accomplished with the aid of Pelouze. He observed that an extract of liver tissue, after being filtered, remained opalescent. When he added alcohol, a white precipitate was formed which he removed and dried. After removing the sugar with alcohol washes, the protein with potash, and the potassium carbonate with acetic acid, the pure glycogen was obtained after a final alcohol wash. The method is still in use.

There have been claims of priority by other authors concerning the isolation and purification of glycogen, but their claims, on careful critical examination of the evidence, have been discarded by Pflüger[21,22] as unfounded and no doubt is left that glycogen was for the first time discovered, isolated and purified by Bernard.

It is certain that the discovery of the glycogenic function of the liver must be considered as a great achievement of "physiological chemistry". The isolation of glycogen was another great success. Bernard's "washed liver" experiment was criticized by his opponents. Figuier, of course, was right when he facetiously credited Bernard with having discovered a "posthumous physiological function". In his "washed liver" experiment, as we know, Bernard was not observing the release of sugar from glycogen as it takes place during life, but the autolysis of glycogen by a different enzymatic system. But the point which was correct in Bernard's experiment is that the source of the glucose is the same in the physiological function and in the *post-mortem* autolysis. It is fortunate that Bernard did not, as his methodology would have required, perform a control experiment with muscle, for instance, for he would also have observed, in "washed muscle" experiments, a liberation of glucose, and he would have had to give up the glycogenic function of the liver.

After Schwann's formulation of the theory of the cells, considering "elementary parts" (our cells) as units of metabolism and the body fluids as their "blastema", providing their nutrients, and after the experimental location of metabolism in cells by Pflüger, the metabolic importance conferred on body fluids by all humoralistic theories was abandoned, and the body fluids were considered as accomplishing a nutritive and protective function of the cells.

In order to appreciate the metabolic revolution introduced by Bernard we must remember that when he followed Magendie's experiments and started his researches on metabolism, the prevalent theory resulting from the views introduced by Liebig and by Dumas and Boussingault taught that the solid parts of animal organisms were composed of nitrogenous substances originating in what Prout had designated as the albuminous part of the food, produced by plants and introduced as such into the animal tissues. No fat or carbohydrate was considered as present in animal tissues, and such components coming from vegetable foods were considered as being directly submitted to combustion in the blood together, according to Liebig, with the carbonaceous derivatives of the slow oxidation taking place in tissues when the repression of these slow oxidations by the vital force (which did not obtain in blood) was reduced by the transformation of parts of this vital force into movement. It is true that Liebig had recognized

References p. 212

the possibility of fat synthesis in animals, but this he considered as an abnormal phenomenon taking place in blood when the supply of oxygen was insufficient. To quote from Liebig's *Animal Chemistry*:

"The production of fat is always a consequence of a deficient supply of oxygen, for oxygen is absolutely indispensable for the dissipation of the excess of carbon in the food. This excess of carbon, deposited in the form of fat, is never seen in the Bedouin or in the Arab of the desert, who exhibits with pride to the traveller his lean, muscular, sinewy limbs, altogether free from fat; but in prisons and jails it appears as puffiness in the inmates, fed, as they are, on a poor and scanty diet; it appears in the sedentary females of oriental countries; and finally, it is produced under the well known conditions of the fattening of domestic animals." (translation of Gregory[23])

Bernard's views on indirect nutrition are in complete opposition with the views generally accepted at the time in the form of what he called the dualist theory of nutrition, formulated by Dumas and Boussingault and according to which the plant is an instrument of reduction and biosynthesis and the animal an apparatus of combustion, analysis and destruction. According to the dualist theory, the nutrients are directly transferred from the plants to the animals where their proximate principles take a place corresponding to their nature.

On the contrary, in Bernard's view, whereas the plants perform photosynthesis, both animals and plants are the seat of similar phenomena of composition and decomposition, in all their elementary parts (cells). Rejecting direct nutrition, *i.e.* the direct transfer of immediate principles, from plants to animals, Bernard believes that nutrition is an indirect phenomenon, as emphasized by his discovery of the process of the synthesis of glycogen in the liver, from different sources of nutrients, providing the blood with a constituent, glucose, which the liver adds to the blood, as he says, as a syringe would do.

5. The concept of the internal environment

As said above (Chapter 6), Schwann's *Mikroskopische Untersuchungen* are composed of three parts. The first and second parts are concerned with the origin of cells and cell differentiation, and the third with cell physiology and metabolism (theory of the cells). In the third part, the concept of a nutritive blastema, *i.e.* a liquid medium in with the cells bathe, plays an important role. Schwann considers, for instance, the sugar solution in which yeast produces alcoholic fermentation as the blastema of the yeast cells. These aspects of the "theory of the cells" had impressed Bernard. Schwann's discoveries as well as his strict reductionist belief, according to which all vital phenomena must be reducible to the laws of physics and chemistry, were far from accepted in France, in contrast with the deep impact they exerted in Germany through Lotze, Mayer, Helmholtz and others.

Schwann's book was never translated into French, and was only made known by a translation, by Lereboulet, of a very unfaithful digest prepared by Müller. The cell theory was rejected by Comte who treated it with the utmost contempt. While everywhere else the word *cell* is used to designate all the forms of parts of organisms, the phrase *cells and fibers*, in which Robin expressed his opposition to the cell theory, is still in common use in French scientific literature nowadays. The attention paid by Bernard to Schwann's views shows, besides other indications, that he was no orthodox adept of positivism. Schwann (born 1810) and Bernard (born 1813) differed little in age and both belonged to the intellectual descendants of Magendie*.

When he discovered the glycogenic function of the liver and developed the concept of indirect nutrition, Bernard was, by a bold conceptual generalization, led to his views on the internal environment and its regulation. Bernard replaced the concept of composition and decomposition—a concept that recurred time and again—by the cellular processes of "chemical creation" (chemical synthesis) and enzymatic degradation, the products of the degradation being used in the process of "chemical creation". The regulation of the internal environment replaced the ancient vitalist views on the harmonies of dynamic permanence.

Blood and other fluids were endowed with an entirely new importance by the concept of the internal environment and of its constancy in higher forms of life, as conceived by Bernard. Another aspect of the concept of dynamic permanence, the internal environment provides the conditions essential for ensuring the function of the cells in an organism too large for them to be in contact with the exterior, and allowing for the conciliation of

* They apparently never met. In 1860, Chr. Retzius, having seen in Schwann's laboratory an animal holder for experimentation on the dog, told Bernard about it and through the intermediation of A. Spring, his colleague in Liège, Schwann was asked by Bernard for a drawing of the apparatus. This was the first personal contact of the two scientists as shown by the following sentence in Schwann's answer (10 December 1860): "Je suis bien charmé d'avoir eu l'occasion d'entrer avec vous dans des relations qui, je le désire vivement, ne cesseront plus"[24]. A year later, Schwann sent his assistant V. Masius to Bernard's laboratory where he stayed from 1861 to 1863. Masius asked Schwann, on behalf of Bernard, for a drawing of the apparatus he used for artificial respiration. This description was sent by Schwann to Bernard with a letter dated 16 April 1862 in which he also expresses his thanks for a gift of curare[24].

References p. 212

the autonomic life of the cells and the integrated whole of the individual.

In a well-documented study, Holmes[25] has related the development of the relation, in the work of Virchow and of Bernard, of the concepts of the cells as units of metabolism as introduced by Schwann, and the fluids of the organism, deprived of their metabolic function by Pflüger's discoveries, and endowed from then on with an integrative function conciliating the autonomous life of cells with the life of the individual. The origins and the development of the concept of the internal environment have been retraced by Grmek[26] who has had access to a great many unpublished manuscripts of Bernard, and from the study of which the present author has drawn extensively in the following passages.

Bernard[27], in the second lecture of his course on general physiology (1854), refers to the topic to which our present chapter is devoted. He states that if nutrition is a word used to designate the faculty of all organisms to feed themselves, it may also be applied to the cells and it consists in the selective attraction exerted by it on its environment, blood or sap of each cell taking up the principles necessary for its constitution and only those required. At the same time decomposition takes place. Therefore blood must be considered as a medium containing all the principles necessary for the maintenance of the life of the cells with which it is in contact. This is clearly inspired, as Grmek points out, by Schwann's "theory of the cells", and implies an "elaboration" of the nutrients. Claude Bernard was preoccupied by the opposition of the solidist and humoralist theories, and he wished to find a solution to the problem that had been dividing the theories of pathology since the opposition between the Greek dogmatists and methodists. In 1855, in his *Leçons de Physiologie Expérimentale*, Bernard[17] defines his concept of "internal secretion" which has nothing to do with the concept of hormonal regulation. What he means is a nutritive secretion playing a rôle in the chemical regulation of blood plasma, the glands introducing into the blood the principles they have elaborated for the nutrition of the tissues. It is clearly stated in a lecture[28] of 16 December 1857,

"As I said it, blood is thus a real internal environment in which all the tissues liberate the products of their decomposition, and in which they find, for the accomplishment of their functions, invariable conditions of temperature, humidity, oxygenation, as well as the nitrogenous, carbohydrate and saline materials without which the organs cannot accomplish their nutrition. Nevertheless, in the nutrition of the organs, we must consider the tissues as active and as acting on the blood to integrate in their structure [*s'approprier*], according to their nature, the different materials of which they are composed." (ref. 28, translation of author)

The concept of the internal environment, a corollary of Bernard's views on indirect nutrition, was in his mind ever since he discovered the glycogenic function of the liver and conceived the idea that glucose, as a constant constituent of blood, was not directly transferred from the food to the blood but was derived from a hepatic reserve, glycogen, synthesized from a number of different nutrients (our gluconeogenesis). In the lecture of 16 December 1857, referred to above[28], he explicitly rejects the old concept of direct assimilation, that is, the introduction of ready-made proximate principles into animal tissues.

"It would not be exact to consider blood as containing all the proximate principles of the organs which would be deposited by a sort of selection in such or such tissue..."

The protective nature of the blood is emphasized, but the conditions provided by the internal medium are already qualified as invariable. Holmes[29] remarks very pertinently that the conditions of the internal medium selected by Bernard do not result from an analysis of its properties but from the concepts used by the physiologists of the time when they enumerated the influences of the external medium on the organism (Tiedemann[30], Béclard[31]).

In his notebook known as the *Cahier Rouge*, Bernard[32] makes precise his position towards humoralism:

"Nothing takes place in the blood itself, but everything in its contact with the tissues." (ref. 32, translation of author)

The cell theory acquired new aspects under the influence of several workers, and particularly in the formulation of Virchow who, amongst other things, replaced the blastema theory of Schwann by the "omnis cellula e cellula". Bernard, in 1860, insists on the autonomy of the cells taking up their nutrients from the blood by their own activity[32]. And again, in 1864, he insists on the notion that the internal environments are the products of the organisms

"It is the product of the organs or apparatus called "of nutrition" (digestion, respiration, secretion) which have no other function than to prepare the general nutritive fluid in which live the organic elements necessary to life, such as the muscular and nervous elements. Moreover there are organs which I have called of *internal secretion* such as the liver, the spleen, the "glandes sanguines" which are only contributing to the composition of the blood by the products they secrete in it. The role of the glycogenic cells is nothing else than to make sugar, one of the components of the blood." (ref. 33, translation by author)

It can be seen that repeated statements of Bernard concerning the internal environment formulate a new kind of humoralism, the humours being the products of the digestive tract and of its accessory glands, as well as of the glands that were to be recognized later as endocrine, and of the respiratory

References p. 212

and excretory mechanisms. These humours are not the seat of the "animal functions", which take place in cells (muscles, nerves, sensory organs), but are consequences of the contact and the interchanges of these cells with the humours.

6. Bernard's views on metabolism

In his *Leçons sur les Phénomènes de la Vie*, Vol. I, published after his death in 1878, Bernard formulated a synthetic view of metabolism in which he combines the "theory of the cells" introduced by Schwann (see Chapter 6), the theory of protoplasm introduced by Huxley (see Chapter 16) and the enzymatic theory of metabolism (see Chapter 13) with his concepts of indirect nutrition and of the internal environment. He links, at the level of cells, the phenomena of composition and decomposition in an original way as he relates the process of composition (renovation) to the process of decomposition (destruction).

"These phenomena are combined in nature, in a sequence which cannot be broken. The two inverse operations of destruction and renovation are absolutely connected and inseparable, in the sense that destruction is the necessary condition of renovation; the acts of destruction are the precursors and the investigators of those by which the parts are restored and reborn, *i.e.* the act of organic renovation." (ref. 34, translation by author)

In the internal environment all the phenomena concur to ensure the integration of the life of the differentiated cells and the continuity of the exchanges of nutrition. Everything is oriented by cells and for cells. The respiratory apparatus introduces oxygen; the digestive apparatus introduces into the internal environment the necessary nutrients, simpler than the food constituents; the circulatory apparatus and the secretory organs ensure the renewal of the internal environment. The nervous system regulates all these machineries and harmonizes them for the benefit of the life of the cells. Enzymes are involved in the phenomena of decomposition taking place in protoplasma, the basic element of life (Huxley's theory), endowed with irritability and mobility. The phenomena of organic destruction correspond to the functions of organisms while the phenomenon of renovation restores the components that have been degraded and consists in a "chemical synthesis or formation of the proximate principles of the living matter...". According to Bernard's views, the nutrition of the cells is accomplished at the expense of the constituents of the internal environment, which are not the constituents of the food but simple molecules resulting from metabolic processes that ensure the preparation of metabolites which will be used for the elaboration of cellular constituents.

These views could not be readily accepted by his contemporaries. They were inspired in part by his endeavour to enforce the unity of life, insisting in a prophetic manner on the simple constituents used by animals as well as plants in biosynthesis. The metabolic theory of Bernard was much more shrewd and penetrating than his contemporaries imagined. It marks an epoch by indicating the dividing line between modern biochemistry and all that went before.

In fact it took a long time completely to substantiate by factual evidence the inductions of Bernard which correspond in many aspects with our present beliefs. We know, as he himself showed, that glucose reaches the blood indirectly by a relay in glycogen; we also know that fatty acids are released into the blood by hydrolysis from reserves of glyceride; and we know that the proteins of foods are introduced into the internal environment in the form of their monomers, the amino acids.

On the other hand, indirect nutrition, according to our present knowledge, is not as general as Bernard thought, and we know that certain nutrients are introduced without modification into the framework of cells as occurs with certain coenzymes. But, in the form in which he formulated it, Bernard's concept of indirect nutrition accomplished a complete revolution in biochemical concepts and put an end to the myth—ever and anon recurring—of direct assimilation, according to which the constituents of the food, either as such, or with slight modifications, were selectively taken into the solid parts to refill, like to like, their own worn-out constituents. Bernard's concept of indirect nutrition foresaw in a penetrating way the developments of our concept of biosynthesis starting from simple materials derived from the complex nutrients (Parts III and IV of this History). As underlined by Canguilhem[35], biosynthesis, called by him "organic creation", was clearly recognized by Bernard as a hereditary structure and not as a succession of states of an isolated system ruled by the Carnot–Clausius principle. We find here a first premonition of what will later be recognized, in one of the most profound contributions of molecular biology, as a coded programme of protein synthesis, a concept of which the origins will be the subject of a later part of this History.

7. Regulation of the internal environment

In 1867, Bernard[36] published a report on the state of French physiology.

References p. 212

In this report, mainly devoted to his own work, the emphasis is shifted from the aspect of protection and nutrition of the cells by the internal environment, towards the concept of its regulation, commanded by the nervous system. This concept states that the nervous system acts at the level of the chemical phenomena taking place "around the organic elements in the internal medium". In his last work, *Leçons sur les Phénomènes de la Vie*, Tome I, published[34] after his death, Bernard formulates the theory according to which all the biological phenomena have a single object, "that of keeping constant the conditions of life in the internal environment". "The organism," he says, "is an equilibrium. As soon as a change occurs in the equilibrium, another takes place to reestablish it." With these words Bernard formulates the concept later called homeostasis by Cannon[37] (in French, homéostasie), and more recently further developed in its cybernetic aspects.

The physiological works of Bernard on secretions, on circulation, on the nervous system and on toxicology have secured for him an important place in the history of physiology. But in spite of, and probably owing to, their great originality, the concepts of indirect nutrition and of the internal environment were not readily accepted. In France, for instance, the metabolic theory of Dumas and Boussingault remained paramount as a consequence of the oppressive influence of the chemists.

It was through such British and American scientists as Bayliss and Starling, Sherrington, J. S. Haldane, Barcroft, Cannon, Henderson and others that the concept of the internal environment was popularized since the beginning of the 20th century[38]. It is paradoxical that the term "homéostasie" was introduced by Lapicque into the French language as a translation of the word "homeostasis" coined by Cannon*.

8. Bernard's idea of experimental medicine

The concept of the life of the cells in a regulated internal environment Bernard[39] himself defines as a different concept of the human body and as the basis of experimental medicine as he conceives it in his *Introduction to the Study of Experimental Medicine*[10,11]. As noted by Canguilhem[35],

* It was during a visit, as lecturer, of L. Lapicque to Cannon's department in Harvard Medical School (Boston, Mass.) in 1931 that this translation was adopted during conversations between Cannon and Lapicque, the role of interpreter being accomplished by the present author, who had previously worked in Lapicque's laboratory and was at the time in Cannon's department.

though the term experimental medicine and the aim contemplated were not new, the content was Bernard's idea, and an entirely new view of medicine was introduced. Here again, Bernard was half a century ahead of his time. The first part of the *Introduction*[10] was immediately received with great favour by the philosophers interested in methodology, and it became a classical text still studied in schools nowadays in France. But it certainly cannot be said that the other parts, those related to medicine did in any perceptible degree influence the clinicians of the time. This did not come as a surprise to Bernard, as his disciple Paul Bert writes:

"He felt the full importance of his own discoveries as foundations for the medical edifice, but he did not share the illusions of those whose eagerness to transfer them to the realm of clinical or therapeutic applications often made him smile. The feeling for distances, which would have discouraged less valiant men, moved him not at all. For strength and perseverance, he did not need the intoxication of illusions. So he, who taught that medicine is or should be a science, showed himself thoroughly skeptical about physicians; and when he talked of them, the shade of Sganarelle* always seemed to pass before him." (translation of Greene[11])

It took time to introduce new concepts into clinical medicine, but after the turn of the century the movement made itself felt and according to the wishes of Bernard when he stated that

"experimental medicine, which is synonym for scientific medicine, can be established only by spreading the scientific spirit more and more among physicians." (translation of Greene[11])

But this is another story.

9. Dynamic state of body constituents

The demonstration by Bernard of the glycogenic function of the liver had eliminated the concept of the inability of animal cells to perform biosynthesis. Many biosyntheses have since been identified in animal cells and, as we shall see later on, a number of aspects as to their metabolic sequences have been unravelled. All cells were considered as being able to accomplish composition by chemical synthesis as well as decomposition. The idea became widespread that cells were able to degrade certain of their molecules and to replenish their quantity by biosynthesis, but the notion persisted of a fixed and stable state of their structural elements as had been proposed by Robin. Contrary to this view of the independent types of exogenous and endogenous metabolism, Borsook and Keighley[40] proposed the concept of a "continuing metabolism" in which exogenous and endogenous con-

* In *Le médecin malgré lui* by Molière.

References p. 212

stituents, food and tissue, take almost equal parts. This view was based on experiments on the rate of metabolism of nitrogen and sulphur.

Later, Schoenheimer[41] proposed the concept of "metabolic regeneration", *i.e.* of the continual release and uptake of substances by cells, from and to a "metabolic pool". In his views on the metabolic "regeneration" of proteins, for instance, Schoenheimer considers that

"The replacement of the amino acids of tissue proteins by their dietary analogues and the transference of nitrogen must involve the rapid breaking and reformation of peptide linkages. As the uptake of nitrogen is a rapid process, it follows that the opening and closing is also a fast reaction. The peptide bonds have to be considered as essential parts of the proteins, and one may conclude that they are rapidly and continually opened and closed in the proteins of normal animals." (ref. 41)

For him,

"the nitrogen excreted by a normal animal in nitrogen equilibrium is *a sample of the pool originating* from the constant and rapid chemical interaction of food and body proteins." (ref. 41)

Schoenheimer states his theory of metabolic regeneration as follows:

"The large and complex molecules and their component units, fatty acids, amino acids and nucleic acids, are constantly involved in rapid chemical reactions. Ester, peptide and other linkages open; the fragments thereby liberated merge with those derived from other large molecules, and with those absorbed from the intestinal tract, to form a metabolic pool of components indistinguishable as to origin. These liberated molecules are again subject to numerous processes. Fatty acids are dehydrogenated, hydrogenated, degraded or elongated, and thereby continually interconverted. While some individual molecules of these acids are completely degraded, other individuals of the same chemical species are steadily formed from entirely different substances, notably from carbohydrate. Similar reactions occur among the split products of the proteins. The free amino acids are deaminated, and the nitrogen liberated is transferred to other previously deaminated molecules to form new amino acids. Part of the pool of newly-formed small molecules constantly re-enters vacant places in the large molecules to restore the fats, the proteins, and the nucleoproteins. Some of the small molecules involved in these regeneration reactions constitute intermediate steps in the formation of excretory products." (ref. 41)

The analogy proposed by Schoenheimer is that of a military regiment*:

"Its size fluctuates only within narrow limits, and it has a well-defined, highly organized structure. On the other hand, the individuals of which it is composed are continually changing. Men join up, are transferred from post to post, are promoted or broken, and ultimately leave after varying lengths of service. The incoming and outgoing streams of men are numerically equal, but they differ in composition. The recruits may be likened to the diet; the retirement and death correspond to excretion." (ref. 41)

While the concept of a metabolic pool, derived from the work of Borsook

---

* *Cf.* the comparison with a monastery in Diderot's *Songe de d'Alembert*, p. 103.

and of Schoenheimer, remains valid, the concept of metabolic regeneration as formulated by Schoenheimer as a chemical exchange between bound and free amino acids, involving breakage of two strong covalent bonds, has not survived modern knowledge of the mechanisms of protein synthesis.

"However, [as stated by Broda[42]] through some mechanism, which is obviously delicately balanced, a steady state in respect to the concentration of circulating protein, and of amino acid is maintained which does not correspond to chemical equilibrium, and therefore not to maximum entropy."

On the other hand there is evidence showing that breaks in DNA may be repaired through the action of DNA ligase.

According to the theory accepted at the present time, most constituents of matter-of-life are in a steady state of rapid flux, whether functional or structural, of simple or of complex constitution. This concept has been extended by isotope techniques to the concept of the continuous "turnover" (synthesis, degradation and replacement) of body constituents, even at constant composition. Here again we find at another level, the theme—recurring time and again—of dynamic permanence.

The topographical situation of the phenomenon of composition and decomposition successively situated at the level of organisms and of cells has been shifted to the sites of the cell's constituent organelles (see Chapter 16, and Vol. 23 of the present Treatise), each of their constituents being characterized by a particular rate of turnover, the dynamic permanence of the organelle being controlled by regulating actions which will be considered later on in this History.

A large number of data has been collected on the cellular and molecular renewal, particularly in the mammalian body. In a recent symposium[43], to which the reader is referred, the dynamic nature of the body constituents was discussed at the level of cells, of organelles and macromolecules.

References p. 212

REFERENCES

1 Ch. Robin and F. Verdeil, *Traité de Chimie Anatomique et Physiologique Normale et Pathologique*, 3 Vols., Paris, 1853.
2 Ch. Robin, *Anatomie et Physiologie Cellulaire*, Paris, 1873.
3 E. P. Cathcart, *The Physiology of Protein Metabolism*, London, 1912 (new ed., 1920).
4 J. Schiller, *Claude Bernard et les Problèmes Scientifiques de son Temps*, Paris, 1967.
5 J. N. Eberle, *Physiologie der Verdauung*, Würzburg, 1834.
6 T. Schwann, *Arch. Anat. Physiol.*, (1836) 90.
7 C. Bernard and L. C. Barreswil, *Compt. Rend.*, 19 (1844) 1284.
8 C. Bernard, *Compt. Rend. Soc. Biol.*, 4 (1877) 244.
9 M. D. Grmek, in *Commentaires sur Dix Grands Livres de la Médecine Française*, Paris, 1968.
10 C. Bernard, *Introduction à l'Étude de la Médecine Expérimentale*, Paris, 1865.
11 C. Bernard, *An Introduction to the Study of Experimental Medicine*, (translated by H. C. Greene), New York, 1949.
12 C. Bernard, *Du Suc Gastrique et de son Rôle dans la Nutrition*, Thèse pour le doctorat en médecine, Paris, 1843.
13 D. W. Wilson, *Popular Sci. Monthly*, 84 (1917) 567.
14 F. G. Young, *Ann. Sci.*, 2 (1937) 47.
15 J. M. D. Olmsted, *Claude Bernard, Physiologist*, New York, 1938.
16 C. Bernard, *Recherches sur une Nouvelle Fonction du Foie Considéré comme Organe Producteur de Matière Sucrée chez l'Homme et les Animaux*, Thèse pour le doctorat en Sciences naturelles, Paris, 1853.
17 C. Bernard, *Leçons de Physiologie Expérimentale Appliquée à la Médecine Faites au Collège de France*, Vol. 1, Paris, 1855.
18 C. Bernard, *Compt. Rend.*, 41 (1855) 461.
19 C. Bernard, *Leçons sur la Physiologie et la Pathologie du Système Nerveux*, Paris, 1858.
20 C. Bernard, *Compt. Rend.*, 44 (1857) 578.
21 E. F. W. Pflüger, *Arch. Ges. Physiol.*, 96 (1903) 1.
22 E. F. W. Pflüger, in Ch. Richet (Ed.), *Dictionnaire de Physiologie*, Vol. 7, article Glycogen, Paris, 1907, p. 228.
23 J. von Liebig, *Animal Chemistry* (translated by W. Gregory), Cambridge, 1842.
24 M. Florkin, *Lettres de Théodore Schwann (1810–1882)*, Liège, 1960.
25 F. L. Holmes, *Bull. Hist. Med.*, 37 (1963) 315.
26 M. D. Grmek, in *Philosophie et Méthodologie Scientifiques de Claude Bernard*, Paris, 1967, p. 117.
27 C. Bernard, *Monit. Hôpitaux*, 2 (1854) 449.
28 C. Bernard, *Leçons sur les Propriétés Physiologiques et les Altérations Pathologiques des Liquides de l'Organisme*, Vol. 1, Paris, 1857.
29 F. L. Holmes, *Arch. Intern. Hist. Sci.*, 16 (1963) 369.
30 F. Tiedemann, *Traité Complet de Physiologie de l'Homme, Traduit par A. J. L. Jourdan*, Paris, 1831.
31 J. Béclard, *Traité Élémentaire de Physiologie Humaine*, Paris, 1855.
32 C. Bernard, *Cahier de Notes 1850–1860, Présenté par M. D. Grmek*, Paris, 1965.
33 C. Bernard, *Rev. Cours Sci.*, 2 (1864) 102.
34 C. Bernard, *Leçons sur les Phénomènes de la Vie Communs aux Animaux et aux Végétaux*, Vol. 1, Paris, 1878.
35 G. Canguilhem, *Études d'Histoire et de Philosophie des Sciences*, Paris, 1968.
36 C. Bernard, *Rapport sur les Progrès et la Marche de la Physiologie Générale en France*, Paris, 1867.

37 W. B. Cannon, *Physiol. Rev.*, 9 (1929) 399.
38 E. H. Olmsted (Mrs. J. M. D. Olmsted), in *Philosophie et Méthodologie Scientifiques de Claude Bernard*, Paris, 1967.
39 C. Bernard, *Principes de Médecine Expérimentale*, Paris, 1947.
40 H. Borsook and G. L. Keighley, *Proc. Roy. Soc. (London), Ser. B*, 118 (1935) 488.
41 R. Schoenheimer, *The Dynamic State of Body Constituents*, Cambridge, Mass., 1946.
42 E. Broda, in *Maria Sklodowska-Curie: Centenary lectures*, Vienna, 1968.
43 I. L. Cameron and J. D. Thrasher (Eds.), *Cellular and Molecular Renewal in the Mammalian Body*, New York and London, 1971.

Chapter 11

From Forces-of-Life to Bioenergetics

1 Vital principles and vital force

Except for those students of the history of the categories of the human mind who are interested in the discursive treatment of nature in primitive stages of knowledge, such concepts as life-as-soul and life-as-action, *vis essentialis*, *nisus formativus*, *physis*, archée, motive principle, *impetum faciens*, vital force, vital power, vital principle, vital spirit, soul, etc., are of little interest and remain outside the domain of biochemistry, a soulless, mechanicist and emergentist science postulating no spiritual principle in the explanations of the changes of the substance or the energy of organisms, in spite of the fact that these entities have been called upon to explain certain aspects of metabolism in proto-biochemical theories. The relation of life and matter, situated at the level of the relation of two abstractions, is also outside the field of biochemistry. Nevertheless, the emergence of bioenergetics was preceded in proto-biochemistry by a recourse to the pseudo-physical concept of "forces".

Forces-of-life have always been present, as we have seen, in the proto-biochemical forms of metabolic theories. The belief that matter-of-life was of a unique kind was related to the notion that it was produced under the stimulus of vital force, the effector of the specific characters of organisms, for those who did not agree on a mechanistic view of life, such as that accepted by the Atomists, and later by Gassendi and Descartes, or Schwann. These specific characters were recognized by Haller in the principles of irritability and sensibility, manifestations of the vital force, disappearing with it at death. The "*anima sive natura*", a spiritual principle that "rules" the matter-of-life, is another name given to the vital force by Stahl, the initiator of the phlogiston theory. Boissier de Sauvages, a Montpellerian disciple of Stahl, calls "nature" the power of the organism to maintain its integrity. Hunter[1] defines what he calls the "living principle" as follows:

References p. 248

Plate 38. Paul Joseph Barthez.

"By the living principle I mean to express that principle which prevents matter from falling into dissolution—resisting heat, cold and putrefaction... I have asserted that life simply is the principle of preservation, preserving it from putrefaction.

"Animal and vegetable substances differ from common matter, in having a power superadded totally different from any other known property of matter, out of which arise various new properties; it cannot arise out of any peculiar modification of matter but appears to be something superadded." (ref. 1)

Hunter rejects the use of the concept of "organization" which was to become popular later.

"Mere composition of matter" [he writes] "does not give life, for the dead body has all the composition it ever had." (ref. 1)

He underlines that the "living principle"

"exists in animal substances devoid of apparent organization and motion... Organization and life do not depend the least on each other. Organization may arise out of living parts, and produce action; but life can never arise out of, or depend on, organization... Organization and life are two different things." (ref. 1)

In France, during the last years of the 18th century, there was a dominant tendency to submit life to the reign of laws, as Newton had done for the non-biological world. Besides those who kept the mechanicist views of Descartes (the animal-machine), two "vitalist" schools became individualized. One, dominant in the "École de Santé" of Montpellier where Th. Bordeu (1722–1776) and P. J. Barthez (1734–1806) were its most illustrious representatives, recognized a unique "vital principle", cause of all life phenomena. Chemistry, according to Barthez[2] can only deal with matter-of-life after it has ceased being alive. The "vital principle" is endowed with a number of forces: force of impulsion, forces of attraction, forces of affinity, motor forces, sensitive forces. Barthez is careful to draw a distinction between the soul and his "vital principle". When man dies, he writes,

"his body goes back to the elements, his vital principle is reunited to that of the universe, and his soul goes back to God, who gave it to him, and who assures its immortality." (ref. 2, translation by author)

The vital principle of Barthez is thus in no way comparable to Stahl's *anima*, and has no spiritualist nature. Inspired by the desire to introduce into the science of biology and of medicine a general principle comparable to gravitation, Barthez postulates a vital principle, a factor common to all organisms, and he does not commit himself to decide whether it is only "a faculty of the living body" or a factor distinct from the body.

Bichat rejects the concept of a unique principle of the organism.

"This principle termed vital by Barthez, *Archeus* by Van Helmont, is an assumption as void

References p. 248

Plate 39. Xavier Bichat.

of truth as to suppose one sole acting principle governing all the phenomena of physics." (ref. 3, translation by author)

For Bichat, the vital properties of the tissues he had individualized consisted of variations of the two Hallerian principles, sensibility and irritability. He thought, as did Haller, that these properties were not of a chemical nature, but on the contrary, opposed to physical and chemical phenomena, as expressed in his famous (and often misunderstood) phrase: "Life is the totality of the forces opposed to death." As death took place, the chemical phenomena dominated.

"A force peculiar to living bodies, maintains their molecules together overcoming the chemical and physical laws to which they would be submitted in their free state: consequently these molecules, after death separate to form new combinations." (ref. 3, translation by author)

Barthez and Bichat are both sometimes called "organo-vitalists", indicating that they find the nature of life in aspects of organization. For Bichat, the chemical compounds present in the living matter are different from those found by the chemist in the dead organisms. It may be stressed here that the concept of the biological actions opposing the realization of the chemical equilibrium reached after death is perfectly sound in the light of modern biochemical knowledge.

Medicus considers man as being composed of matter, vital force (a single unifying agency) and soul[4]. The idea of the vital force was submitted to a critical study by Reil (1759–1813), a disciple of Kant, who concluded that the vital force was the expression of the organization of the living parts, resulting from the mixture and form of the matter and forces involved. Reil lists a number of forces: physical force, vegetative force, animal force, etc.[5].

Von Kielmeyer[6] studied the distribution of "organic forces" in the world, and he concluded that sensibility dominates in higher animal forms while irritability is predominant in the lower forms. He also considers the force of secretion and the force of propulsion. All these forces were recombined in a single vital force acting not at the level of organization, but inside matter, by Brandis[7] (1795), as well as by Hufeland[8] (1762–1836) (concept of life as immanent to matter).

As we have seen (Chapters 6 and 7), vital force is considered by Liebig as of the nature of a physical force, acting as controller and repressor of the other forces. His theory is related to that of Paracelsus or Van Helmont whose *Archeus* was "the chemist in the body", ruling the chemical forces. As well as Paracelsus or Van Helmont, Liebig, in spite of his vitalism adheres to a chemical theory of life, and that is why his writings exerted

References p. 248

such an appeal. He is, in contradistinction with organo-vitalists such as Barthez or Bichat, a chemo-vitalist.

Far from presenting any internal contradiction, as is sometimes said, the views of Liebig are, on the contrary, incorporated in a very coherent structure. The vital force he describes as a physical force; the operator or repressor of chemical forces accounts for his preoccupation to explain dynamic permanence. Without making the same force the origin of movement, he could not account for the command of muscles by the will. On the other hand, by attributing to a derepression from vital force the oxidation of azotized molecules in muscles, the source of heat and of substances submitted to combustion and production of heat, he avoided introducing the heat production in a domain ruled by the vital force, as suggested by the opponents of the theory of combustion introduced by Lavoisier.

2. The opposition towards vital force

Chevreul was among the first who paid little heed to a neutralization of chemical forces by vital forces as understood by Barthez and by Bichat. In his widely read classic on organic analysis[9], published in 1824, he writes:

"Whenever we consider the assimilation of organic matter by an organism, we should not decide that in the phenomenon, the vital force chooses what is corresponding to this living organism, and that the assimilated matter is fixed by virtue of a force which differs from affinity; because the fact that, in organisms, compounds are produced which differ from those belonging to the inorganic world, is no sufficient reason to call for a cause other than affinity: indeed, if some compounds are specific of organized matter, others, such as carbonic acid, are inorganic compounds as well, and besides we recognize that the same elements, in different circumstances, are assembled as different compounds; for example, if you mix iron deutoxide with carbon at room temperature, as long as the conditions remain the same, the mixture remains unchanged, but if you heat it enough, you will obtain iron, carbon dioxide and carbon monoxide. No less than heat, electrical effects may modify the nature of combinations and decompositions taking place between substances in contact; for instance, copper, which does not decompose water, is oxidized in water under the influence of voltaic electricity. If we adopt the attitude of avoiding to consider, in molecular actions, affinity as an absolute force, independently of the circumstances in which the atoms between which it acts are situated, we will say that, if the matter assimilated to an organism produces compounds different from those found in raw material, it is because this matter is placed in circumstances characteristic of the organization. But the links binding the atoms remain those of cohesion and affinity; thus we do see that when life has ceased to animate it, a living being still keeps its form, its appearance, during a period of time the length of which depends on the degree of its protection from the contact with exterior influences. If the atoms of this living being had obeyed forces other than affinity to constitute molecules; and if these molecules had obeyed forces other than cohesion to form aggregates, as soon as the elements would cease being submitted to the forces which united them in life, they would dissociate and become free, or part of new compounds which would certainly be of inorganic nature." (ref. 9, translation by author)

In 1837, Chevreul again expressed his opposition to vital force, basing his reasoning on an analogy with an observation he made in the field of dye chemistry. The text we refer to, first appeared in the *Mémoires de l'Académie des Sciences*[10] as an appendix to the sixth paper of the series published by Chevreul on the chemistry of dyeing, one of his favourite subjects, even more attractive to him since he became in 1824 the head of the dyeing laboratory of the "Manufacture Royale des Gobelins". This text was published again during the same year in the *Journal des Savants*[11]. Chevreul observed that Prussian blue, after being fixed on silk or cotton and exposed to light *in vacuo*, bleaches by losing cyanogen and becoming protocyanide. This observation explained the bleaching—under the influence of sunlight—of a cloth dyed with Prussian blue and the recovery of the blue colour in the dark. Chevreul was impressed with the analogy of the course of these events with the course of events in the organisms when they excrete a portion of their substance in one step and absorb oxygen in a following one.

"Suppose" [he writes] "that an organism would contain Prussian blue in a liquid acting as sap or as blood and that this fluid would penetrate in an organ submitted to the action of light which would reduce this colour principle into cyanogen and protocyanide: let us suppose that there be an evolution of cyanogen, and after that an absorption of oxygen, and that this oxygen being carried together with the protocyanide to the site of organs not submitted to the action of light, a formation of Prussian blue and iron peroxide would take place there: I feel confident to say now that such phenomena as the exhalation of cyanogen and the bleaching of Prussian blue in the organ receiving light and the return of the colour of the fluid, following an introduction of oxygen and a discontinuation of the influence of sunlight, would be considered by those who ignore the properties of Prussian blue as an effect of a vital force, while those who are aware of these properties, when coming across this colouring matter in the liquid part of a living creature, and observing the phenomena described above, would soon find an explanation of the colouration and decolouration of the fluid, without any recourse to a vital force." (refs. 10, 11, translation by author)

In the same text, Chevreul states the principles of his biochemical methodology, in two propositions: (*1*) that its basis is the definition of the "chemical species" composing the matter of an organism, a definition which includes not only the elementary composition and the physical, chemical and organoleptic properties of these molecular species but also their molecular arrangements; and (*2*) that no fruitful study of the phenomena taking place in the organisms will be possible without a knowledge of the molecular species as defined above. These concepts, formulated in a famous, widely read and classical book, exerted a lasting influence among French physiological chemists.

While accepting as of an entirely physical and chemical nature such

References p. 248

functions as digestion, respiration, excretion etc., Chevreul was careful to state repeatedly that he was not ready to consider that growth and regulation in an organism belonged to the same category of phenomena as those of metabolism.

In Germany, where *Naturphilosophie* was reigning, the reaction against vital forces was inaugurated by a contribution of Schwann to Müller's *Handbuch der Physiologie*[12], a book which introduced into Germany the experimental method of Magendie. Until his death, Müller remained a convinced vitalist. Recourse to experimentation was for him (as it had been for Bichat) a means of studying the effects of the vital force peculiar to each organ. Restricted in his chemical and physical background, he was to detach himself progressively from physiology to devote himself entirely to comparative morphology, in which field he acquired fame. From the beginning of his career as a researcher, on the contrary, Schwann took a completely different position, which inaugurates the quantitative period of physiology.

Müller's *Handbuch* was in no way a work of mere compilation, for he critically examined all the notions that he printed. Repeating the experiments of others, imagining new ones, opening avenues not yet explored, this treatise is a work unique in its conception as in its realization. In the section entrusted to him, Schwann enriched Müller's treatise with the results of extensive work, and contributed numerous new notions: the structure of voluntary muscle, the existence of a special capillary wall, the muscular contractibility of arteries, the regeneration of severed nerves, the structure of elastic tissue, etc. This treatise also contains an account of a study clearly showing the innovating tendency of Schwann, the first experiments of which can be dated on the basis of his laboratory notebooks as 16 April 1835. In these texts, Schwann envisaged various experiments in which it would be possible to subject the physiological properties of an organ or of a tissue to physical measurement. One such method involved measuring the secretion of a gland. But it was the muscle which seemed to him likely to furnish the most rewarding results. He planned to measure for different loads the length of a muscle contracted by the action of the same stimulus; or, further to compare the intensity of the contraction with that of the stimulus. He accomplished the experiment by means of the "muscular balance", and in a sense established the first tension-length diagram.

It is difficult for us to appreciate the sensation produced in physiological circles by this simple experiment.

"It was for the first time, as Du Bois-Reymond has underlined, that someone examined an eminently vital force as a physical phenomenon and that the laws of its action were quantitatively expressed." (Fredericq[13], translation by author)

In a milieu in which the idealistic philosophy and the theories of Fichte and Hegel were still dominant, the "Fundamental Versuch" came as a revelation and constituted the point of departure for a new physiology.

Soon a campaign of great vigour was raging against vital force,

"a campaign where the first shot was fired by Schwann in 1839 in the same book that contained his interpretation of the cellular origin of the animal organism." (Temkin[14])

The intellectual intentions that led Schwann to conceive of the cell theory are in relation with his views on vital force. Vitalism was the prevailing theory in the laboratory of Müller, the master of Schwann. The mechanicist and unitarian antagonism of Schwann towards this intellectual attitude had already manifested itself in his studies on muscles, on digestion and on fermentation. This tendency to introduce a more exact mode of explanation than that in terms of the vital force then in vogue, was to find its culmination in the formulation of the cell theory.

Schwann has himself defined his attitude towards the vital force, such as was accepted by his master, Johannes Müller, author of the notion of the proper energy of tissues:

"A simple force different from matter, as it is supposed, the vital force would form the organism in the same way as an architect constructs a building according to a plan, but a plan of which he is not conscious. Furthermore, it would give to all our tissues that which is called their proper energy, *i.e.* the properties which distinguish living tissues from dead tissues: muscles would owe it their contractibility, nerves their irritability, glands their secretory function. Here, in a word, is the doctrine of the vitalist school. Never was I able to conceive the existence of a simple force which would itself change its mode of action in order to realize an idea, without however possessing the characteristic attributes of intelligent beings. I have always preferred to seek in the Creator rather than in the created the cause of the finality to which the whole of nature evidently bears witness; and I have also always rejected as illusory the explanation of vital phenomena as conceived by the vitalist school. I laid down as a principle that these phenomena must be explained in the same way as those of inert nature." (ref. 15, translation by author)

Schwann aimed at replacing teleogical explanation by physical explanation. For him, the phenomena of life were not produced by a force acting according to an idea, but by forces acting blindly and with necessity, as in physics. Individual finality itself, such as was considered to be observed in each organism, was determined by the same manner as in inert matter, the explanation depending entirely upon the characteristics of matter and the blind forces with which it had been created by an infinitely intelligent being. He found the confirmation of this view already pre-existing in his mind, in

References p. 248

the notion of the uniformity of the texture and the growth of animals and plants, such as he had developed in his cellular theory.

"The uniformity of this development demonstrated that it is the same force which everywhere unites molecules into cells, and that this force could be nothing but that of molecules or atoms: the fundamental phenomenon of life therefore had to have its *raison d'être* in the properties of atoms." (ref. 15, translation by author)

Schwann's text had such a lasting impact that it is worth quoting at length:

"We set out, therefore, with the supposition that an organized body is not produced by a fundamental power which is guided in its operation by a definite idea, but is developed according to blind laws of necessity, by powers which, like those of inorganic nature, are established by the very existence of matter. As the elementary materials of organic nature are not different from those of the inorganic kingdom, the source of the organic phenomena can only reside in another combination of these materials, whether it be in a peculiar mode of union of the elementary atoms to form atoms of the second order, or in the arrangement of these conglomerate molecules when forming either the separate morphological elementary atoms to form atoms of the second order, or in the arrangement of these conglomerate molecules when forming either the separate morphological elementary parts of organisms, or an entire organism. We have here to do with the latter question solely, whether the cause of organic phenomena lies in the whole organism, or in its separate elementary parts. If this question can be answered, a further inquiry still remains as to whether the organism or its elementary parts possess this power through the peculiar mode of combination of the conglomerate molecules, or through the mode in which the elementary atoms are united into conglomerate molecules.

"We may, then, form the two following ideas of the cause of organic phenomena, such as growth, etc. First, that the cause resides in the totality of the organism. By the combination of the molecules into a systematic whole, such as the organism is in every stage of its development, a power is engendered, which enables such an organism to take up fresh material from without, and appropriate it either to the formation of new elementary parts, or to the growth of those already present. Here, therefore, the cause of the growth of the elementary parts resides in the totality of the organism. The other mode of explanation is, that growth does not ensue from a power resident in the entire organism, but that each separate elementary part is possessed of an independent power, an independent life, so to speak; in other words, the molecules in each separate elementary part [our cell] are so combined as to set free a power by which it is capable of attracting new molecules, and so increasing, and the whole organism subsists only by means of the reciprocal action* of the single elementary parts. So that here the single elementary parts only exert an active influence on nutrition, and totality of the organism may indeed be a condition, but is not in this view a cause.

"In order to determine which of these two views is the correct one, we must summon to our aid the results of the previous investigation. We have seen that all organized bodies are composed of essentially similar parts, namely, of cells; that these cells are formed and grow in accordance with essentially similar laws; and, therefore, that these processes must, in every instance, be produced by the same powers. Now, if we find that some of these elementary parts, not differing from the others, are capable of separating themselves from the organism, and pursuing an independent growth, we may thence conclude that each of the other elementary

* "Reciprocal action" must here be taken in its widest sense, as implying the preparation of material by one elementary part, which another requires for its own nutrition.

parts, each cell, is already possessed of power to take up fresh molecules and grow; and that, therefore, every elementary part possesses a power of its own, an independent life, by means of which it would be enabled to develop itself independently, if the relations which it bore to external parts were but similar to those in which it stands in the organism. The ova of animals afford us examples of such independent cells, growing apart from the organism.

"It may indeed, be said of the ova of higher animals, that after impregnation the ovum is essentially different from the other cells of the organism; that by impregnation there is something conveyed to the ovum, which is more to it than an external condition for vitality, more than nutrient matter; and that it might thereby have first received its peculiar vitality, and therefore that nothing can be inferred from it with respect to the other cells. But this fails in application to those classes which consist only of female individuals, as well as with the spores of the lower plants, and, besides, in the inferior plants any given cell may be separated from the plant, and then grow alone. So that here are whole plants consisting of cells, which can be positively proved to have independent vitality. Now, as all cells grow according to the same laws and consequently the cause of growth cannot in one case lie in the cell, and in another in the whole organism; and since it may be further proved that some cells, which do not differ from the rest in their mode of growth, are developed independently, we must ascribe to all cells an independent vitality, that is, such combinations of molecules as occur in any single cell, are capable of setting free the power by which it is enabled to take up fresh molecules. The cause of nutrition and growth resides not in the organism as a whole, but in the separate elementary parts—the cells. The failure of growth in the case of any particular cell, when separated from an organized body, is as slight an objection to this theory, as it is an objection against the independent vitality of a bee, that it cannot continue long in existence after being separated from its swarm. The manifestation of the power which resides in the cell depends upon conditions to which it is subject only when in connexion with the whole (organism).

"The question, then, as to the fundamental power of organized bodies resolves itself into that of the fundamental powers of the individual cells." (ref. 16, translation of Smith)

It may be seen that Schwann introduces here into the structure of the matter of cells the concept of molecule as defined by the chemists and that he develops a completely mechanistic view of cell chemistry, paving the way for cell biochemistry.

As underlined by Canguilhem, the philosophical sin of vitalism was to propose the insertion of the living organism into a physical medium to the laws of which he constitutes an exception.

"There cannot be any empire in an empire, or else there is no more empire, either as container or as contents." (ref. 17)

This was the thought which led Schwann to his opposition to vital force in which he denounced an unacceptable exception to the laws of God's empire.

Schwann, as Descartes had done for the animal machine, concentrated the finality of nature into the action of God when creating the atoms, all the aspects of nature resulting automatically from the properties of the atoms.

As Temkin[14] pointed out, the solution given by Schwann to the problem of finality, transferring it from biology to the universe and from the "vital

References p. 248

force" to the Creator has found extension in several fields of thought. Lotze[18] was inspired by it in his famous study on the nature of life. Schwann's text, cited above, is the origin of the mechanicist materialism of which R. Mayer, Brücke, Du Bois-Reymond, Helmholtz and Carl Ludwig were the main representatives. It put aside the organo-vitalism of Bichat as well as the chemo-vitalism of Liebig and replaced it by a complete reduction to a chemical or physical basis.

It would be easy to demonstrate by many quotations how highly praised was Schwann's book in German physiological circles. In a letter to Mayer (14 December 1842) the physiologist Griesinger advises his friend to follow the example of Schwann:

"Die Ausbildung und Durchführung einer rein physikalischen Ansicht der Lebenprocesse halte ich für die Aufgabe der Physiologie unserer Zeit. Es muss dir bekannt sein, welche glänzende Beiträge zu solcher z.B. Schwann geliefert hat." (ref. 19)

In his answer, dated 16 December 1842, Mayer[19] writes:

"Durch seine mikroskopischen und mechanischen Versuche hat Schwann sich allerdings ein bleibendes glänzende Verdienst erworben."*

In the field of forces-of-life, Schwann in his extreme reductionist attitude only recognized the existence of the physical properties of the molecules of matter and he had no need of other forces to account for the living machine. For him the force-of-life is a special property of certain matters. Schwann also teaches that the cells are composed of molecules of special structure, and these theories recall the concept of Epicurus relating that quality is the result of a spatial arrangement of atoms.

3. The principle of the conservation of energy

It was around the middle of the 19th century that the concept of the conservation of energy fully emerged, and that it was first recognized that the disappearance of a "force" is always accompanied by the appearance of another "force". Rosen[20] stresses the relation of this development with the context of the industrial revolution and its needs. He rightly underlines the fact that machine technology, rational business calculation as well as the principle of accounting paved the way for an effective accountancy for all forms of energy. And this is what the concept of the conservation of energy did afford. Watt's steam engine was the starting point of a complete change

* Mayer refers to the "theory of the cells" and also to the "fundamental Versuch" of Schwann.

of perspective. Cornwall, the first principal sphere of exploitation of Watt's engine, on account of the necessity for pumping in the mines, became the place where

"the history of accountancy of energy in economic terms begins effectively with the ingenious method developed by Watt to charge royalties on the engines which he and his partner installed." (Rosen[20])

The firm of Boulton and Watt charged a royalty of one-third of the difference between the cost of coal used by the engine and the feed of the horse doing the work of pumping. It was in the circle of the French engineers that Sadi Carnot first developed the concept of the conservation of energy, but his notes remained unknown until 1878, almost half a century after he died of cholera in 1832.

It was from the consideration of forces-of-life that the concept of the indestructibility of energy first emerged in the mind of Mayer.

There are several reviews and books discussing the intricate history of the principle of the conservation of energy in greater detail than space permits here, and we are primarily interested in the aspects derived from the consideration of forces-of-life as it obtained for Mayer and Helmholtz. It has been said that the principle of the conservation of energy offers an example of simultaneous discovery by several independent workers; but, as noted by Elkana:

"The various discoverers discovered different things because they asked different questions. Mayer discovered the indestructibility of forces of nature; Helmholtz discovered that the sum of the various kinds of force is a constant and that all must have the dimension of mv^2; Joule discovered the mutual convertibility of heat and "mechanical powers" as a result of which Thomson established the dynamical theory of heat on a firm foundation. Finally Claussius, Thomson and Rankine showed the equivalence of these various results and it becomes clear that one result is derivable from the other *i.e.* that all of them had actually discovered the same thing." (ref. 21)

Mayer, who can be credited with the formulation of the indestructibility of forces in nature, as he explicitly states[22], followed Schwann's ways towards bringing physiology into the domain of physical and chemical sciences. In 1889, Preyer[19] published a series of letters of Mayer to his friend Griesinger together with the answers, and this constitutes a most interesting source of information.

Mayer had served as a doctor on a Dutch ship, and during a sea voyage to Java, while performing venesections, he noticed that the venous blood of the sailors of his ship was redder in Java than it had been in Europe. He interpreted this observation by stating that in a tropical climate metabolic

References p. 248

Plate 40. Julius Robert Mayer.

activities were less intense, because of the decrease in the body's need for heat production. However, the body also produces work which in turn produces heat. He thought that there must be a definite relationship between the heat produced in the body, the work performed by the body in a given period, and the heat produced by this work. It was known that work produces heat, but Mayer introduced the concept of a certain amount of heat corresponding to a certain amount of work as well as the concept of the indestructibility of "forces".

In a letter to Griesinger (16 June 1844), he describes the flash of insight which led him to bring all the aspects of life under general explanatory principles.

"This theory was by no means hatched in a study. After having zealously and persistently occupied myself with the physiology of the blood during my journey to the East Indies, the observation of the altered physical condition of our crew in the tropics and the process of acclimatization gave me much additional food for thought. The forms of disease as well as the character of the blood in particular continuously directed my thoughts above all to the production of animal heat by the respiratory process. Now, if one wants to achieve clarity concerning the physiological matters, a knowledge of physics is indispensable unless one prefers to treat the matter metaphysically, to which I have an infinite aversion. Consequently, I devoted myself to physics and developed such an intense interest in the problem that I cared little about that distant part of the world. Some may laugh at me on this account, but I preferred to remain on board ship where at certain times I felt as if inspired, nor can I recall anything similar either before or after. Several flashes of thought that struck me—it was on the wharf at Sorabaja—were immediately and eagerly pursued, and in turn led to new subjects. Those times are past, but the deliberate, reflective testing of that which then suddenly emerged within me has taught me that it is a truth which is not only subjectively felt but can also be objectively proven..." (ref. 19, translation of Rosen[20])

When he came home in February 1841, Mayer developed his idea, and he sent to Poggendorff, for his *Annalen*, a paper in which he had calculated the mechanical equivalent of heat. Poggendorff, showing his chronic blindness to flashes of genius, as he had done previously with the paper of Kützing on fermentation, considered Mayer's paper to be devoid of interest and lost the manuscript. Mayer sent another copy to Liebig who published it in his *Annalen* in 1842[23].

Griesinger[19], in a letter to Mayer dated 4 December 1842, tells of the pleasure he felt in reading his friend's paper in Liebig's *Annalen*, and he points out to him that his theory may have some relation to what he has just read in Liebig's *Animal Chemistry*. He refers to the passage in which Liebig calculates, according to the experiments of Despretz, the amount of heat theoretically liberated by the oxidation of the carbon daily converted into CO_2 in the human body, and concludes that it is sufficient to explain the

References p. 248

constant temperature of the body. This text had already been published by Liebig in his *Annalen*[24], and this had led Mayer to send his manuscript to Liebig. In his answer to Griesinger (5–6 December 1842)[19] Mayer tells his friend that any apprehension he might have had concerning priority had vanished when he read the section on *The phenomena of motion in the animal organism* (Part III of *Animal Chemistry*).

"It is nothing but a new patch on an old garment; instead of going ahead with the necessary radicalism, he mixes up new ideas with old errors and falls into sheer mistakes." (translation by author)

In another letter (22 June 1844) he refers to Liebig's

"Hypothesenconglomerat über die Lebenskraft, aus dem die Wissenschaft nichts machen kann." (ref. 19)

We have already referred to the views of Liebig on the vital force, to which he clung to the end of his life, to his metabolic theory (Chapter 7) and to his theory of fermentation (Chapter 6). In the section of *Animal Chemistry* referred to by Mayer, Liebig brings additional traits to his definition of vital force and its implications in the phenomena of motion in the animal organism. We may recall that he considered movement not as a result of chemical phenomena but as a transformation of vital force, as a consequence of which a corresponding amount of what he called slow oxidation took place (see Chapter 7). This kind of relation of vital force with movement as understood by Liebig certainly had nothing in common with Mayer's views, unless one is ready to accept that Liebig discovered the equivalence of the vital force with heat and work.

In 1845, Mayer published his main contribution to bioenergetics[22], in a booklet in which he refutes Liebig's theory of movement on the basis of Schwann's experiments on muscle, and expresses the notion of the transformation of chemical force into mechanical force by muscles. He concludes that the total quantity of heat produced in the body is in a constant ratio to physiological combustions, *immediately as well as indirectly* (by mechanical or other form as well as directly by heat).

Helmholtz[25], a member of Schwann's circle in Berlin, who had been influenced by Schwann's strict reductionism, started from the consideration of animal heat the studies which led him to show mathematically the truth of the principle of the conservation of energy, and to give the standard formulation of this principle in mathematical terms.

Mayer pointed out that what he had been looking for was the numerical relationships according to which heat, motion and kinetic energy can be converted one into the other. As he had realized, his conception afforded the possibility of bringing all the manifestations of life (the so-called "forces" produced by the vital force principle, etc.) under general explanatory principles. On 14 June 1844, Mayer wrote to Griesinger:

"In addition it occurs to me that perhaps you expect that I will indicate a more specific application to physiology. Here one must proceed very slowly and carefully. The one that comes most immediately to mind is a consideration of animal metabolism. For some time a logical instinct had led physiologists to the axiomatic proposition: No action without metabolism. From the physical side this thesis is already definitely expressed in my theory. However, in physiology it is a question of "How and what and when and where?" You will admit that until now a solution of these questions was unthinkable. To achieve this, the physical theory I have developed is in my opinion a necessary requisite." (ref. 19, translation of Rosen[20])

In a text written about 1850, and published only after his death, Mayer states:

"In the living animal body carbon and hydrogen are oxidized and heat and motive power produced in return. Now applied directly to physiology, the mechanical equivalent of heat shows that the process of oxidation is the physical condition for the organism's ability to do mechanical work, and at the same time it indicates the numerical relations between consumption and performance." (translation of Rosen[20])

4. Theories of energy metabolism

The theories of energy metabolism in animals developed as a consequence of the formulation of the principle of the conservation of energy.

By the great impact of Liebig's *Animal Chemistry*, the mutual action between the elements of the food and oxygen as the source of animal heat had became standard. But it was only after the formulation of the principle of the conservation of energy that the sources of animal heat could be unified under the concept of the metabolism of food substances. According to their theoretical views, Lavoisier, as well as Dulong and as Despretz, only took into account the oxidation of carbon and of hydrogen, when reckoning their heat accounts. Important progress came with the formulation, in 1840, of the principle of Hess, according to which the amount of heat produced in a chemical process is independent of the intermediary steps through which the system passes from the original state to the final state.

Besides the error of Liebig, who rejected the application of the Hess principle in the field of life, another error to which Mayer[22] called attention

References p. 248

Plate 41. Max Rubner.

consisted in considering heat as the only way through which the energy produced by metabolism leaves the organism, not realizing that heat can be taken as such a measure only if no other form of energy is produced.

Rubner[26], a disciple of Voit, had observed that at the end of the fasting period, after the last traces of fat are consumed, an animal metabolizes an amount of protein corresponding to the same amount of energy obtained from fat in the previous days, an observation that confirmed the concept of the prevailing importance of the energy content of food.

Progress in energy metabolism came from the method introduced by Voit for the calculation of the amount of protein metabolized by the determination of the nitrogen excreted. This had been suggested by Liebig in his *Animal Chemistry*. This subject was first experimentally studied by Bidder and Schmidt[27]. They found that almost all the nitrogen given to cats and dogs in the form of meat was recovered in the urine and the faeces.

This was expressed by them as follows:

"Almost all the nitrogen of protein and collagen is split from its combination and carries with it enough carbon, hydrogen, and oxygen to form urea; the remaining part, containing five-sixths of the total heat value of the protein, undergoes oxidation to carbon dioxide and water which are eliminated in the respiration, the calorifacient function having been fulfilled." (ref. 27, translation of Lusk[28])

These results were a source of controversy until Voit[29] established what he called the "nitrogen equilibrium".

Voit fed a dog for 58 days with 29 kg of meat containing 986 g of nitrogen. In the excreta for the same period he found 982.8 g of nitrogen (943.7 g in the urine, 39.1 g in the faeces). As the difference between the nitrogen eaten and the nitrogen excreted was so small, it appeared that the way of excretion of nitrogen was through the urine and faeces and that other ways of excretion were negligible.

But one point remained obscure. Is the nitrogen of air introduced into organic compounds in the body? Is any protein nitrogen given off as nitrogen or ammonia gas? How much is lost in sweat or through the nails and hair? According to Lavoisier, nitrogen gas had no function in respiration. Regnault and Reiset[30] found that animals under a bell-jar sometimes absorb nitrogen and sometimes give it off, but always in small amounts, within experimental error. Bachl[31] observed that a rabbit does not expire more than traces of ammonia in the breath. With regard to the cutaneous excretion of nitrogenous material, widely varying results had been obtained. Favre[32] had found 0.044 g of urea for 1000 ml of perspiration (0.02 g of

References p. 248

nitrogen) whereas Funke[33] found a much larger amount, 1.55 g of urea for 1000 ml of perspiration (0.87 g of nitrogen). Argutinsky[34] found 0.363 and 0.410 g of urea in 225 and 330 ml of perspiration respectively. Benedict[35] showed that in a resting man the cutaneous excretion may amount to 0.071 g nitrogen *per day*, in a man at moderate work to 0.13 g *per hour* and at hard work for four hours to 0.22 g *per hour*. In order to find the possible loss of nitrogen through hair, nails and epidermis, Moleschott[36] cut the hair and nails of several men once a month. The outgrowth of hair daily amounted to 0.029 g of nitrogen and that of nails to 0.0007 g of nitrogen. Here again the amounts are negligible.

Leo[37] devised experiments to determine whether an animal extracts nitrogen originating from another source than the inspired air, and he found that a maximum of 0.5% of the nitrogen of the diet may appear in the expired air. Although several sources of error had been detected in these experiments, Krogh[38], in very well-controlled experiments, concluded that atmospheric nitrogen plays no part in respiration, the whole of the ingested nitrogen appearing in the urine and faeces.

Voit had established the method of determining the amount of protein metabolized by a determination of urinary urea nitrogen. As meat protein in general contains 16% nitrogen, 1 g of urinary nitrogen corresponds to 6.25 g of metabolized protein. Voit suggested to Pettenkofer the construction of an apparatus in which a measurement could be performed of the total carbon excretion whether in respiration, in urine or in faeces. This expensive piece of apparatus was built in Munich with the financial help of King Maximilian II of Bavaria. With its help Pettenkofer and Voit[39] studied the carbon dioxide of expired air, and during the same period, the excretion of carbon and nitrogen. An experiment they performed is summarized by Lusk[28] as follows:

"The man was allowed a small quantity of Liebig's extract of beef, as the experimenters did not at that time realize the very slight discomfort usually entailed by total abstinence from food. As Liebig's extract has no nutritive value, its effect has been counted out in the following description.

"The subject, on entering the living-room of the apparatus, weighed 71.090 kg, and he drank during the day 1.0548 liters of water, making a total body weight of 72.1448 kg. Twenty-four hours later he weighed 70.160 kg and his excreta had amounted to 0.7383 kg carbon dioxide, 0.8289 kg water from lung and skin, and 1.1975 kg of urine. The final body weight plus all the excreta amounted to 72.9247 kg. A total body weight of 72.1448 kg was converted into a body weight plus excreta amounting to 72.9247 kg. The difference is due to oxygen absorbed. The difference of 0.7799 kg represents the amount of oxygen needed to convert the body substance lost into the excretory produces obtained. The tabular statement reads:

MAN—STARVATION

	kg		*kg*
Weight at start	71.090	Weight at end	70.160
Water drunk	1.0548	Carbon dioxide	0.7383
		Water in respiration	0.8289
Oxygen absorbed	0.7799	Urine	1.1975
	72.9247 kg		72.9247 kg

"The analysis of the urine showed 12.51 g of nitrogen and 8.25 g of carbon. A calculation gives the amount of carbon in the respiration as 201.3 g. If we neglect the faeces as being too small in starvation to influence the results, we find that the total carbon elimination for twenty-four hours was 209.55 g and the total nitrogen 12.51. In the Liebig extract ingested there was 2.44 g of carbon and 1.18 g of nitrogen which must be deducted from the above in order to obtain the strict loss of carbon and nitrogen from the body during the period of starvation. These values are:

C	207.11 g
N	11.33 g

"These two figures enabled Pettenkofer and Von Voit to calculate what substances had burned in the body. As every g of nitrogen in the excreta is approximately represented by the destruction of 6.25 g of meat protein, the amount of such protein destroyed by the man was 70.81 g. It has been found that for every g of nitrogen present in meat protein there are 3.28 g of carbon. It is therefore easy to estimate that destruction of protein represented by 11.33 g of nitrogen involved the elimination of 37.16 g of carbon. Now, the man eliminated 207.11 g of total carbon, from which this protein carbon may be deducted, leaving as residue 169.95 g, which must have been originated from a source other than protein. The possible sources are two in number—carbohydrates and fats. In starvation no carbohydrates are ingested and their supply in the form of reserve glycogen is usually counted as being negligible in such experiments as these. The only other source from which the 169.95 g of extra carbon could have been derived is fat, and as fat contains 76.52 per cent of carbon, a destruction of 222.1 g of fat may be calculated. This fasting man therefore destroyed:

Protein	70.81 g
Fat	222.1 g

"That such metabolism actually did take place was further indicated by the comparison of the amount of oxygen needed for the destruction of the above constituents, and the amount of oxygen absorption as determined by the experiment.

"From the constituents of the protein and fat destroyed, Pettenkofer and Von Voit deducted the constituents of the urine, which contains part of the C and H belonging to protein. The balance of the carbon and hydrogen was fit for oxidation to carbon dioxide and water. Their calculation may thus be presented:

	Weight in grams		
	C	*H*	*O*
Composition of the protein burned	37.16	5.8	17.1
Composition of fat burned	169.95	25.7	25.1
Total C, H and O metabolized	207.11	31.5	42.2
Deduct quantity in the urine	8.2	2.0	7.6
Balance available for respiratory CO_2 and H_2O	198.9	29.5	34.6

References p. 248

Plate 42. Max Pettenkofer.

Oxygen required	530.4	235.7	
Total O required for the formation of CO_2 and H_2O			766.1
Less O in the protein and fat			34.6
Oxygen actually required			731.5
Oxygen absorption as determined			779.9
Difference			48.4

"We may reach the same result by using the most modern figures for the oxygen requirement in the metabolism of the food-stuffs. We now know that to burn 100 g of meat protein requires 133.43 g of oxygen, and to burn 100 g of fat requires 288.5 g, and to burn 100 g of starch requires 118.5 g. This being true, there are required:

	Oxygen (g)
For 70.81 g protein	94.44
For 222.1 g fat	639.55
Total required	733.99
Oxygen absorption as found	779.9
Difference	45.91

"Had carbohydrates burned, less oxygen would have been needed, since carbohydrates contain a larger proportion of oxygen than fats. Had the extra 169.95 g of carbon been due to the combustion of starch (or glycogen), 382 g would have burned, requiring 452.7 g of oxygen instead of 639.5 g for fat. Pettenkofer and Von Voit found in the amount of oxygen absorption a confirmation of their belief that the fasting organism supports itself by the combustion of its own protein and fat.

"It is apparent from this discussion that *the quantity of oxygen needed in metabolism depends upon the chemical composition of the material that burns in the organism*, and also that the relation between the amount of oxygen absorbed and carbon dioxide excreted depends on the same factor. Regnault and Reiset frequently observed that this latter relationship was variable. The ratio of the *volume* of carbon dioxide expired to the *volume* of oxygen inspired during the same time is called *respiratory quotient* (R.Q.). When carbohydrates burn, the R.Q. is unity; that is, for every hundred volumes of carbon dioxide excreted a hundred volumes of oxygen are absorbed. When protein burns the quotient is

$$\frac{\text{Vol. } CO_2}{\text{Vol. } O_2} = \frac{78.1}{100} \text{ or } 0.781$$

and when fat burns the quotient is 0.71. Pettenkofer and Voit calculated that the respiratory quotient in their fasting man was 0.69. This indicated a combustion of fat in the organism." (ref. 28)

In his necrology of Pettenkofer, Voit[40] describes the enthusiasm of the inventors of the apparatus:

"Imagine our sensations as the picture of the remarkable processes of the metabolism unrolled before our eyes, and a mass of new facts became known to us! We found that in starvation protein and fat alone were burned, that during work more fat was burned, and that less fat

References p. 248

was consumed during rest, especially during sleep; that the carnivorous dog could maintain himself on an exclusive protein diet, and if to such a protein diet fat were added, the fat was almost entirely deposited in the body; that carbohydrates, on the contrary, were burned no matter how much was given, and that they, like the fat of the food, protected the body from fat loss, although more carbohydrates than fat had to be given to effect this purpose; that the metabolism in the body was not proportional to the combustibility of the substances outside the body, but that protein, which burns with difficulty outside, metabolizes with the greatest ease, then carbohydrates, while fat, which readily burns outside, is the most difficultly combustible in the organism." (ref. 40, translation of Lusk[28])

As we have stated in another chapter, Voit showed that muscle work did not increase protein metabolism and that the metabolism was not proportional to the supply of oxygen. The amount of metabolism determines the amount of oxygen absorbed.

According to Mayer's principle, energy cannot arise from nothing. Where it is active it must have been potential elsewhere. Is the energy appearing in the body as mechanical work and electric currents, all of which may be measured as heat, entirely derived from the metabolism?

Bischoff and Voit[41], in 1860, still obtained the values of the potential energy of the nutrients by calculation from the amounts of CO_2 and H_2O. This method was recognized as being unacceptable, and it appeared necessary directly to measure the heat of combustion of the nutrients themselves. Favre and Silbermann[42], who first methodically performed these experiments, determined the heat of combustion of several organic compounds, among them fatty acids, but the components of the diet which, at the time, were supposed to be used as such by the tissues (proteins, fats, sugars), were not studied by them. Berthelot[43], in 1865, established the conditions and methods for obtaining valuable data of this kind. Voit[44], who had brought back from England a Thomson calorimeter, compiled in 1866 a table of values he used in his course of lectures. From the table he calculated that the fasting man studied by Pettenkofer and Voit[45] (see above) had a metabolism corresponding to 2.5 million small calories.

In 1873, Pettenkofer and Voit[45] estimated that 100 g of fat were the physiological equivalent of 175 g of starch. This concept of *isodynamy* was the subject of the investigations of a student of Voit, Schürmann. As Schürmann died before the completion of his work, Rubner followed on the same track and determined the equivalent quantities of the different foodstuffs:

100 g fat = 232 g starch = 234 g cane sugar = 243 g dried meat

But doubts still remained concerning the theory of combustion as stated by Lavoisier. Information about this lack of agreement can be obtained by reading the *Leçons sur la Chaleur Animale* of Claude Bernard[46]. In this book, Bernard considers a function of the higher organism, the physiological function of heat production which appears as a primary property of these organisms, maintaining their temperature between given limits, and regulated with a definite objective. The theory that had generally been accepted since Lavoisier was the theory of direct combustion, according to which oxygen burns the carbon and the hydrogen of the nutrients, an operation resulting in the production of CO_2 and water, and in a production of heat resulting from this direct combustion. Pflüger demonstrated that the operations of metabolism take place in the cells, and much documentation showed that all organs produce heat. Muscle contraction was an example of these producers of heat. But, says Bernard, why should the heat generated by a muscle be considered as resulting from a direct combustion of carbon and hydrogen?

To answer this query Rubner performed a famous series of experiments. He started from the principle that the contribution to body heat of the "burning" of starch and fat was principally the same as the value determined in the calorimeter, as in both cases the initial and final products (CO_2, H_2O) were the same. But when protein is metabolized in the body, the products are partially lost by urine and faeces: this corresponds to chemical energy, or latent heat, lost by the body, and this amount must be taken into account.

It had been customary since the time of Stohmann[47], to deduct the heat value of urea from the heat value of protein in order to obtain the fuel value of protein for the organism. But in experiments performed by Pettenkofer and Voit in 1866 (see above) it had been noticed that in dogs fed with meat, or starved, there was a much larger output of carbon in the urine than that corresponding to the urea present. In starvation, there was an increase of other nitrogenous excreta (creatinine, uric acid, etc.). Rubner understood that it was the heat value of the actual urinary constituents which should be subtracted from the heat values of proteins in order to determine the fuel value of these proteins.

After actually burning the dry residue of urine, Rubner obtained the calorific values of urine (Table I).

It was also necessary to know the heat value of the faeces. Rubner determined the amount of heat produced from 1 g of ash-free faeces after an

References p. 248

TABLE I

Calorific values of urine

	C/N	*Calories from 1 g*	*Calorific value of 1 g N*
Urea	0.429	2.523	5.41
Urine after feeding protein	0.532	2.706	5.69
Urine after feeding meat	0.610	2.954	7.46
Urine in starvation	0.728	3.101	8.49

ingestion of meat and found 6.127 calories; and after an ingestion of washed meat (protein) he found 6.852 calories. Burning 1 g of beef in the calorimeter, Rubner obtained 5.345 calories. If from 100 g of meat, 2.7 g appear as faeces with a calorific value of 6.127 calories per g, there is a loss of $6.127 \times 2.7 = 16.83$ calories.

If from 100 g of meat containing 15.4 g of nitrogen 15.16 g of this nitrogen appears in urine, and the urine has a calorific value of 7.46 calories per g of nitrogen, the energy loss in the urine would be $7.46 \times 15.16 = 112.94$ calories. From the values above, Rubner calculates the fuel value of dry muscle (Table II).

TABLE II

The fuel value of dry muscle

			Calories
100 g muscle			534.5
Waste	urine	112.94	129.77
	faeces	16.83	
Fuel value of 100 g of dry muscle			404.73

In drying the muscle, as well as in dissolving the urea and other components of the urine, Rubner estimates the loss of heat as follows

for the drying of protein	2.688
for dissolving the urea, etc.	1.989
	4.677 calories

Subtracting the value from the fuel value obtained above for 100 g of dry meat, gives $404.73 - 4.67 = 400.06$ calories. For the same material the calorimeter shows a heat value of 534.5 calories. The conclusion is that 74.9% of this (400.06) is available in the organism, while ±25%, the remainder, goes to waste.

Rubner estimated the calorific values of proteins in different physiological conditions (Table III). From these values, it is possible to calculate the heat value of the metabolized proteins from the amount of N in the excreta.

TABLE III

Calorific values of proteins in different physiological conditions

	Calories yielded by 100 g of protein in the body	*Heat values corresponding to 1 g of N in excreta*
After ingestion of washed meat (protein)	442.4	26.66
After ingestion of whole meat	400.05	25.98
In starvation	384.2	24.98

For pig's fat, Rubner calculated that the heat value per g was 9.423 calories. Since fat contains 76.5% of carbon, 12.3 calories are liberated in the body for every g of carbon eliminated in the expired air. Using these values, Rubner recalculated the amount of heat liberated by the fasting man of Pettenkofer and Von Voit (see above). He multiplied by 24.98 the nitrogen excreted and by 12.3 the carbon of the metabolized fat and obtained the total heat value for the period:

Heat from protein (11.33 g N × 24.98)	283 cal
Heat from fat (169.95 × 12.3)	2091 cal
Total heat value of metabolism	2374 cal

In the course of this work, Rubner formulated what has been called the "surface law", according to which, in organisms of different body sizes, the metabolic rate is proportional to their respective surface areas. The historical origins of the concept are worth recalling. As pointed out by Kleiber[48], it had indubitable finalist origins. Sarrus and Rameaux[49], who first formulated the concept in 1839, considered that it was the aim of nature to make the rate of heat production of large and small animals in proportion to their respective surface areas or to the two-third power of their body weights.

Rubner deduced from his data the simple rule that fasting homeotherms

References p. 248

produce daily approximately 1000 calories of heat per square meter of body surface. Still working in the Munich laboratory, Rubner carefully verified a series of standard values for the average fuel value of nutrients[50].

1 g of protein	4.1 cal
1 g of fat	9.3 cal
1 g of carbohydrate	4.1 cal

5. Metabolism as the only source of energy in the animal body

After having moved to Marburg, Rubner built a calorimeter which could accurately measure the heat produced by a dog in 24 h. As a respiration apparatus was attached to the chamber of the calorimeter, the metabolism could be calculated by the method of Pettenkofer and Voit, and from the metabolism the heat production could be estimated. With the help of this apparatus, Rubner[51] collected a number of data leaving no doubt as to the validity of the theory according to which *the metabolism is the only source of energy in the body*, as Table IV shows.

TABLE IV

Comparison of estimated heat from metabolism with heat actually produced

Food	*Number of days*[a]	*Heat calculated from metabolism*	*Heat directly determined per day*	*Difference (in percentage)*
Starvation	5	1296.3	1305.2	−1.42
	2	1091.2	1056.6	
Fat	5	1510.1	1498.3	−0.97
Fat and meat	8	2492.4	2488.0	
	12	3985.4	3958.4	
Meat	6	2249.8	2276.9	−0.42
	7	4780.8	4769.3	+0.43

[a] Spent in calorimeter.

On the basis of this classical table of data the theory of the metabolism as the energy source of the organism became standard. The whole series of works on energy metabolism of the Munich and Marburg schools is therefore in continuity with the theory of metabolism considered as a phenomenon of combustion, as first proposed by Lavoisier.

The principle of the apparatus devised by Pettenkofer, Voit and Rubner, as well as by Regnault and Reiset, was used for the determination of metabolism in clinics. This is one of the aspects of the utility of calorimetry for measuring any form of energy and this is why chemical energy became ordinarily expressed as calories.

Animals convert chemical energy supplied as nutrients to other varieties of chemical energy (body substance, milk, etc.), or to other forms of energy (work, heat, etc.). In an animal without any of the forms of net synthesis of body substance (growth) or body products (milk, eggs, etc.) the overall intensity of all anabolic and catabolic processes (metabolic rate) corresponds to the catabolic rate, as there is no *net* anabolism. If there is no net change in the total pool of energy-rich bonds and if there is no work performed, the heat production equals the catabolic rate.

The trend of development of the technique of metabolism measurements by the indirect method made a detour through the technique of mining-rescue apparatus (see Florkin[15]).

In 1896 Laulanié[52] performed respiration calorimetric measurements on a number of animal species, fasted or fed (guinea-pigs, rabbits, ducks, dogs), and he found that the mean caloric equivalent per litre of oxygen consumed, when measured directly, was 4.75 kcal, and the equivalent calculated from the carbon and nitrogen balance was 4.71 kcal per litre on the average. It is sometimes said that Rubner, as well as Laulanié, showed, in the work referred to above, the validity of the law of conservation of energy as applied to living organisms. As correctly pointed out by Kleiber[48], what they confirmed for organisms is the law of constant heat sums, as formulated by Hess in 1840.

Rubner's work did not deal with one of the major aspects of the first law of thermodynamics: the relation of heat to work. That mechanical work in animals is in agreement with the first law was demonstrated by Atwater and Rose in 1899[53] and by Atwater and Benedict[54] a few years later. They used a respiration calorimeter to measure the chemical changes and the production of heat in men working on bicycle ergometers. The human work efficiency (ratio of work performed to increase in chemical energy spent) ranged in general from 13 to 21%.

"For practical purposes, [concluded the authors] we have therefore warranted in assuming that the law [of energy conservation] obtains in general in the living organism as indeed there is every reason *a priori* to believe that it must." (ref. 54)

On the basis of the law of Hess it was possible to calculate the heat production of chemical processes.

References p. 248

6. Organisms considered as heat engines

As stated above, in the whole series of contributions of Pettenkofer, Voit and Rubner, the main idea was to demonstrate the theory according to which animal heat was a product of metabolism and quantitatively equalled the chemical energy of the molecules metabolized. The conditions were chosen to avoid the accomplishment of work, in order not to divert any amount of it, as heat was considered to be the source of work. On this point it will suffice to read in Bernard's *Leçons sur la Chaleur Animale*, p. 400:

"Si l'on appelle chaleur ou *ration d'entretien* celle qui échauffe l'animal à un degré constant, et *ration d'activité* celle qui se transforme en travail, il faut reconnaître qu'il n'y a entre elles deux aucun antagonisme, aucune réciprocité nécessaire." (ref. 46)

Other famous physiologists of the time, such as Pflüger and Engelmann, shared the view according to which mechanical work resulted from the production of heat, a part of which was transformed into other forms of energy. The physiologists were conscious of the difficulty resulting from a consideration of the second law of thermodynamics, and they tried to avoid it by a number of assumptions.

The views according to which work was derived from the heat produced in organisms by the metabolism of chemical compounds was based by Engelmann on the postulation of the existence in the organism of a cycle of Carnot. In order to produce work from heat in a closed cycle, there must be a fall of temperature. In fact Engelmann[55] proposed the idea of the dark and clear disks of muscles functioning as heat engines.

The theory adopted by Pflüger[56] to account for metabolism and the energy produced by it, consists in the assumption that CO_2 is formed inside big molecules and produces an explosion of these molecules, the consequence of which is the production of vibrations and of a temporary local high increase of temperature (± 10000 °C) which he considers as the source of the mechanical work. Such fantastic theories could not resist criticism based on thermodynamic calculations, as presented by Gautier[57] and by Fick[58]. Therefore, the view according to which the chemical energy of metabolized substances directly produced mechanical or chemical energy became current, accompanied by a declaration of agnosticism concerning the processes of this transformation. It was this situation that prevailed at the end of the 19th century.

7. Bioenergetics

The whole of the classical physiological work on "energy metabolism" taught nothing about the "go" of living things. But in other scientific circles, the concept of the conservation of energy and of the interchangeability of its forms grew into the science of thermodynamics, concerned with the principles of laws governing material transformations. While the organism was, at the beginning of the 19th century, believed to consume vital force, it is considered nowadays as an open system, and as the seat, not only of a flux of matter, but also of a flux of energy. In 1923, the publication of Lewis and Randall's *Thermodynamics and the Free Energy of the Chemical Substances*[59] brought to the world of biochemists a knowledge of the concepts introduced into thermodynamics by Gibbs and of the two functions *energy* (the variation of which determines the possible total heat exchange) and *entropy* (the variation of which determines the possible performance of work).

It was Kluyver[60] who, in 1924, first pointed out that endergonic reactions can proceed only when suitably coupled with exergonic reactions, and he raised the problem of such energy coupling. But, as shown in later chapters of this work, no satisfactory solution could be proposed until the demonstration by Warburg in the years 1937–1938 of the formation of ATP (which was known as an energy-rich compound for several years) in a coupling to the first oxido-reduction of glycolysis (enzymatic oxidation of phosphoglyceraldehyde) and the demonstration by Meyerhof of the same sort of coupling with the second energy-yielding reaction of glycolysis (phosphoglycerate to phosphoenolpyruvate).

As we shall see, it was only during the period 1937–1941 that oxidative phosphorylation was specifically coupled to respiration. The source of energy, therefore, became the oxido-reductions derived from the transfer of electrons provided by nutrients, and the cells were recognized as isothermal chemical engines. In a general way, the energy transformed in organisms originates from reactions in which phosphorylation is coupled, as we shall see in Part III or IV of this History, with the use either of a chemical compound or of light. What became of heat as the dominating concept of the old metabolic theories, in that picture? Animal heat, in this context, becomes a waste from the successive steps of the workings of the isothermal chemical engines of cells.

References p. 248

Plate 43. Albert Jan Kluyver.

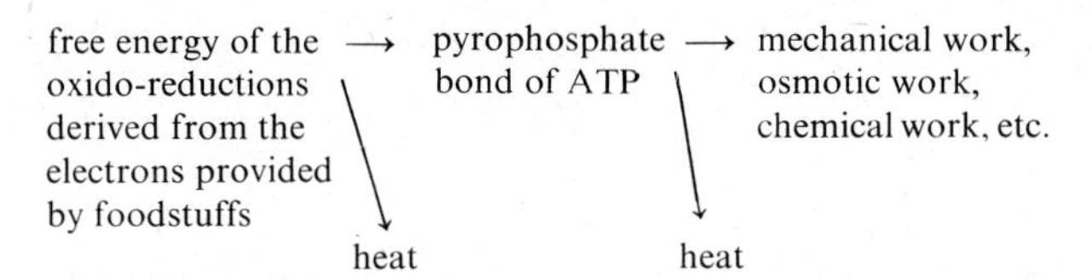

That this scheme also applies to plant cells is one of the teachings of the modern theory of mitochondrial biochemistry. What makes the green plant a specialized biosynthesizer of carbohydrates from CO_2 absorbed from the air and the water from the soil, depends on the nature and biochemistry of chloroplasts, the seats of photosynthesis, containing chlorophyll.

Before the definition of the concepts of energy and of entropy, the systems of the Greek philosophers, for instance, saw in the "innate heat" the power accounting for the production of heat without any obvious source of heat, explaining, according to Aristotle, all the functions of organisms. Galen saw in heat, the source of which was in the heart, not only the motor of digestion but also of the evolution of the *pneuma* from the blood.

After Lavoisier, heat, though recognized as produced by combustion resulting from respiration, still remained the motor of the actions of life. Modern biochemists have realized, as will be related in later chapters, that redox reactions are the ultimate source of all the energy of life, and that in all cells a large part of the oxidation–reduction energy is converted into phosphate-bond energy, the universal "go" of life, the remaining part being dispersed as waste in the form of heat. Heat became a waste product and lost the overvalorization which had originated in the mind of man by the tragic opposition of the cold of death and of the warmth of life.

References p. 248

REFERENCES

1 J. Hunter, *A Treatise on the Blood*, in J. Palmer (Ed.), *Collected Works of John Hunter*, London, 1837.
2 P. J. Barthez, *Nouveaux Éléments de la Science de l'Homme*, 2 Vols., 2nd ed., Paris, 1806.
3 X. Bichat, *Anatomie Générale*, Paris, 1802 (translation by C. Coffyn).
4 F. K. Medicus, *Von der Lebenskraft*, Mannheim, 1774.
5 J. C. Reil, *Arch. Physiol.*, 1795.
6 C. F. von Kielmeyer, *Über die Verhältnisse der organischen Kräfte untereinander in der Reihe der verschiedenen Organisationen, die Gesetze und Folgen dieser Verhältnisse*, Stuttgart, 1793.
7 J. D. Brandis, *Versuch über die Lebenskraft*, Hannover, 1795.
8 C. W. Hufeland, *Die Kunst, das menschliche Leben zu verlängern*, Jena, 1797.
9 M. E. Chevreul, *Considérations Générales sur l'Analyse Organique et ses Applications*, Paris, 1824.
10 M. E. Chevreul, *Mém. Acad. Sci.*, 18 (1837).
11 M. E. Chevreul, *Journal des Savants*, Nov. 1837.
12 J. Müller, *Handbuch der Physiologie des Menschen für Vorlesungen*, Coblenz, 1834–1838.
13 L. Fredericq, *Théodore Schwann: Sa Vie et ses Travaux*, Liège, 1884.
14 O. Temkin, *Bull. Hist. Med.*, 20 (1946) 322.
15 M. Florkin, *Naissance et Déviation de la Théorie Cellulaire dans l'Oeuvre de Théodore Schwann*, Paris, 1960.
16 T. Schwann, *Microscopical Researches into the Accordance in the Structure and Growth of Animals and Plants* (translated from the German by H. Smith), Sydenham Society, London, 1857.
17 G. Canguilhem, *La Connaissance de la Vie*, Paris, 1952.
18 H. Lotze, in *Wagners Handwörterbuch der Physiologie*, Vol. 1, Braunschweig, 1842, p. 9.
19 W. Preyer, *Robert von Mayer und die Erhaltung der Energie*, Berlin, 1889.
20 G. Rosen, in C. McC. Brooks and P. F. Cranefield (Eds.), *The Historical Development of Physiological Thought*, New York, 1959, p. 243.
21 Y. Elkana, *Arch. Intern. Hist. Sci.*, 23 (1970) 31.
22 J. R. Mayer, *Die organische Bewegung in ihrem Zusammenhange mit dem Stoffwechsel*, Heilbronn, 1845.
23 R. Mayer, *Ann. Chem. Pharm.*, 42 (1842) 233.
24 J. von Liebig, *Ann. Chem. Pharm.*, 41 (1842) 189.
25 H. von Helmholtz, *Über die Erhaltung der Kraft*, Berlin, 1847.
26 M. Rubner, *Z. Biol.*, 17 (1881) 214.
27 F. H. Bidder and C. Schmidt, *Die Verdauungssafte und der Stoffwechsel*, Milan, 1852.
28 C. Lusk, *The Elements of the Science of Nutrition*, Philadelphia and London, 1917.
29 C. von Voit, *Physiologische Untersuchungen*, Augsburg, 1857.
30 V. Regnault and J. Reiset, *Ann. Chim. Phys.*, 26 (1849) 299.
31 M. Bachl, *Z. Biol.*, 5 (1869) 61.
32 P. A. Favre, *Compt. Rend.*, 35 (1852) 721.
33 O. Funke, *Untersuchungen zur Naturlehre der Menschen und der Thiere*, 4 (1858) 36.
34 P. Argutinsky, *Arch. Ges. Physiol.*, 46 (1890) 594.
35 F. G. Benedict, *J. Biol. Chem.*, 1 (1905–1906) 263.
36 J. Moleschott, *Untersuchungen zur Naturlehre*, 12 (1868) 187.
37 H. Leo, *Arch. Ges. Physiol.*, 26 (1881) 218.
38 A. Krogh, *Sk. Arch. Physiol.*, 10 (1900) 353.
39 M. Pettenkofer and C. von Voit, *Z. Biol.*, 2 (1866) 478.
40 C. von Voit, *Z. Biol.*, 41 (1901) 1.

41 C. G. Bischoff and C. von Voit, *Die Gesetze der Ernährung des Fleischfressers*, 1860.
42 P. A. Favre and J. T. Silbermann, *Ann. Chim. Phys.*, 34 (1852) 357.
43 M. Berthelot, *Ann. Chim. Phys.*, 6, 4e Ser. (1865) 444.
44 C. von Voit, *Münchener Med. Wochschr.*, 49 (1902) 233.
45 M. Pettenkofer and C. von Voit, *Z. Biol.*, 9 (1873) 534.
46 C. Bernard, *Leçons sur la Chaleur Animale sur les Effets de la Chaleur et sur la Fièvre*, Paris, 1876.
47 F. A. C. Stohmann, *J. Prakt. Chem.*, 31 (1885) 273.
48 M. Kleiber, *The Fire of Life*, New York, 1961.
49 M. M. Sarrus and Rameaux, *Bull. Acad. Roy. Med.*, 3 (1839) 1094.
50 M. Rubner, *Biol.*, 42 (1901) 261.
51 M. Rubner, *Z. Biol.*, 30 (1894) 73.
52 F. Laulanié, *Arch. Physiol.*, (1896) 572.
53 W. O. Atwater and E. B. Rosa, *U.S. Dept. Agr. Off. Expt. Sta. Bull.*, 63 (1899).
54 W. O. Atwater and F. G. Benedict, *U.S. Dept. Agr. Off. Expt. Sta. Bull.*, 136 (1903) 1.
55 T. W. Engelmann, *Über den Ursprung der Muskelkraft*, 2nd ed., Leipzig, 1893.
56 E. F. W. Pflüger, *Arch. Ges. Physiol.*, 10 (1875) 641.
57 A. Gautier, *Cours de Chimie*, Vol. 3, *Chimie Biologique*, 2nd ed., Paris, 1897.
58 A. Fick, *Arch. Ges. Physiol.*, 53 (1893) 606; 54 (1893) 313.
59 G. N. Lewis and M. Randall, *Thermodynamics and the Free Energy of Chemical Substances*, New York, 1923.
60 A. J. Kluyver, *Chem. Weekblad*, 21 (1924) 266 (English translation in A. F. Kamp, J. W. M. la Rivière and W. Verhoeven (Eds.), *Albert Jan Kluyver; his Life and Work*, Amsterdam, 1959).

Chapter 12

Life Banned from Organic Chemistry

1. Wöhler's synthesis of urea

About the middle of the 19th century it was a general belief that organic substances, composing the matter of organisms, could not be formed outside them. The synthesis of urea by Wöhler[1] in 1828 did not introduce any rupture with the common belief; urea was considered to be an exception. As Schwann[2] states in a lecture he delivered in Louvain in 1841,

"Urea contains the same elements, and in the same proportions, as ammonium cyanate and this salt is easily transformed into urea, bringing an example of artificial formation of an organic substance." (translation by author)

Schwann, the initiator of the antivitalistic campaign in Germany, did not consider that the synthesis of urea by Wöhler could be used as an antivitalist weapon. Liebig[3], in 1840, spoke of urea "as the first organic compound artifically produced", and Dumas[4] immediately after its realization mentioned "Wöhler's brilliant discovery of the artificial production of urea". When he wrote about his accomplishment to Berzelius in February 1828, Wöhler said

"I can make urea without the necessity of a kidney, or even of an animal, whether man or dog. Ammonium cyanate is urea."... "Can this artificial production of urea be regarded as an example of the formation of an organic substance from inorganic bodies? It is remarkable that for the preparation of cyanic acid (and also of ammonia) an organic substance is always originally necessary." (ref. 5, translation of Hopkins[12])

It is obvious, from these lines, that Wöhler himself did not claim that he had accomplished a complete synthesis of urea in his paper entitled "Über Künstliche Bildung des Harnstoffs"[1]. Several authors shared the same view[6,7]. But these authors, as well as Wöhler, were wrong, as potassium cyanide (converted into cyanate by fusing in lead oxide) had been obtained by Scheele (1783) from potassium carbonate, graphite and ammonia gas, and ammonia by Priestley by reduction of nitric acid, synthesized from its

References p. 263

Plate 44. Friedrich Wöhler.

elements by Cavendish (1785) (see Partington[8]). Both Berthelot[9] and Meyer[10] recognized that Wöhler had really accomplished a complete synthesis of urea from inorganic bodies. All the authors mentioned rightly considered[11] that Wöhler's synthesis of urea could not be counted as identical with the biological pathway of urea synthesis, and therefore they considered it as artificial. As noted by Hopkins[12], Wöhler was himself more interested in the point also emphasized in Berzelius's answer to his letter than to any vitalistic position.

"It is an altogether striking circumstance", Berzelius had remarked, "that the character of a salt is so completely absent when the acid unites with ammonia, a circumstance which will certainly be very enlightening for future theories." (translation of Hopkins[12])

As has been stressed by Brooke[13], placed in its true light, Wöhler's main achievement lies in its implication of a "structural" programme for organic chemistry and for organic synthesis.

As also remarked by Hopkins[12], the textbooks published during the decade following Wöhler's synthesis of urea made little of the antivitalist significance emphasized by more modern treatises, written after the breach in vitalist organic chemistry had been widened by later discoveries. It was certainly one of the events at the start of the development of the synthesizing activities of the chemists, and in that context one of the pioneer works of a development which was finally to expel "vital force" from organic chemistry.

2. The nucleus theory of Laurent

The significance of the synthesis of urea, for most of the chemists, was an aspect of the discussion of the theories of radicals[8], as imagined by Berzelius who claimed that in organic matter the atoms enter into combination two by two (binary combination). For example, he represents sodium sulphate as Na_2O+SO_3, while in his opinion the organic compounds, instead of being binary, were tertiary, quaternary, etc. In the theory of Berzelius, the inorganic compounds are composed of an electropositive radical and an electronegative radical. These two kinds of radical he also considers according to his theory as existing in organic substances composed of electronegative radicals such as cyanogen and of electropositive radicals such as ethyl. The theory was abandoned when organic combinations containing only two atomic species, such as the olefins, were discovered. The concept of the relation of the properties of the molecules with the spatial arrangements of atoms, which played such a major part in the development of organic

References p. 263

Plate 45. Auguste Laurent.

chemistry, had its first origins in an observation of Gay-Lussac that if wax be treated with chlorine, hydrochloric acid is formed as a result of a substitution of chlorine for hydrogen atoms of the wax. Dumas made the same observations with other substances, and as the substances in which chlorine had been substituted for hydrogen did not show any considerable changes in their properties, it appeared that these properties depend less on the nature of the constituent atoms, but rather on their arrangement in the molecules. This concept led Laurent to formulate his theory of the "nucleus" in his doctoral thesis of 1837[8]. He considered that the molecules of organic substances are formed in a manner similar to the development of crystals. As there is in crystals a primitive form of a certain mathematical character, and as other layers can be deposited on it, Laurent proposed recognizing the existence, in the organic molecules, of a primitive mathematical form, the nucleus, on the surface of which atoms may be deposited. It is the nucleus that imposes the main properties on the molecule. Let us take a prism with 16 limiting surfaces. At each angle a carbon atom is located (32 C) and between two angles a hydrogen atom (32 H). At the level of the two pointed extremities of the prism, two molecules of water are situated. The formula of this molecule is $C_{32}H_{32} \cdot 2H_2O$, and its elementary formula is $C_{32}H_{34}O_2$. According to this theory acetic acid is $C_4H_4+O_4$, C_4H_4 being the nucleus ethene. The nucleus theory of Laurent not only played a part in the development of the theory of types but it seems to be the origin of the parallel of the cell formation with a crystallization, as it appears in the "theory of the cells" of Schwann. The *Mikroskopische Untersuchungen* of Schwann was written in 1838 and published in 1839. Laurent's thesis had appeared in 1837. That Schwann's attention had been seriously attracted by this thesis is substantiated by the fact that in his course[2] on general anatomy given at Louvain in 1841, Schwann devotes a part of one of his lectures, delivered to medical students, to the nucleus theory of Laurent. This is certainly no full proof of filiation but is surely an element in favour of not discarding it. (On the comparison of living matter with crystals, repeatedly found in the history of biology, see Hall[14].)

3. The theory of chemical types

The "theory of chemical types" has played a very important rôle in the history of organic chemistry as a science[8]. It consists of the concept of the existence of definite inorganic "types" from which all organic compounds

References p. 263

Plate 46. Charles Gerhardt.

could be derived. A first type proposed by Hofmann in 1849 was ammonia. A second type, water, was proposed by Williamson in 1850. Adding hydrogen and hydrochloric acid, Gerhardt, in 1853, considered that the organic compounds were derived from the four types: H_2O, NH_3, H_2 and HCl, to which Kekulé added the type CH_4 in 1856. Gerhardt included the radicals in his theory of types, in the form of Laurent's "residues" as he defined them, as fragments that do not exist in the free state, but are apt to substitute themselves for elements or for other radicals. Gerhardt also assimilated and generalized the concept of "chemical function" introduced by Peligot and Dumas in 1836. As a consequence of the application of types to hydrocarbons, Gerhardt also introduced the notion of the homologous series[8,15].

4. Structural formulas of organic compounds

By his work and by the application of the discoveries of his friend Laurent, Gerhardt[16] appears as the principal initiator of the structural formulas of organic compounds. He introduced the concept in 1848. He considers a molecule as an edifice, a unique system formed by the assembly in a determinate but unknown order, of infinitely small particles called atoms, either of the same kind or of different kinds. For him the structural formula allows the provisions of reactivities but, being another victim of the positivist theories then prevailing in France under the leadership of Auguste Comte, Gerhardt did not dare to believe that the formulas could represent the real spatial arrangements of atoms. Many chemists could not accept the idea that the structural formula of a compound could be derived from chemical reactions. That the structure is not radically changed by reactions was accepted by Frankland, by Williamson and by Odling, a concept confirmed later by the "principle of inertia" formulated by Van 't Hoff[17]. It was Butlerow (1828–1886) who first claimed that the structural formulas derived from the reactivities represent the spatial positions of the atoms⋆.

To base the representation of molecular structures on the knowledge of reactions may not be the only possible approach: the fact that it has been preferred by organic chemists to other possible ways of formulation has made possible the development of the biochemistry of intermediary metabolism consisting of the identification and seriation of the reactions taking place at the level of the chemical components of cells.

⋆ On the aspects of the beginnings of organic chemistry mentioned above, the reader is referred to Partington[8].

References p. 263

Plate 47. August Kekulé.

Reverting to structural formulas we must stress that the kind of graphology finally adopted by the organic chemists was one of the conditions of the development in studies relating to the sequences in metabolic pathways and was one of the basic foundations of the great strides made by scientific biochemistry. The history of the use of graphic formulas goes back to the

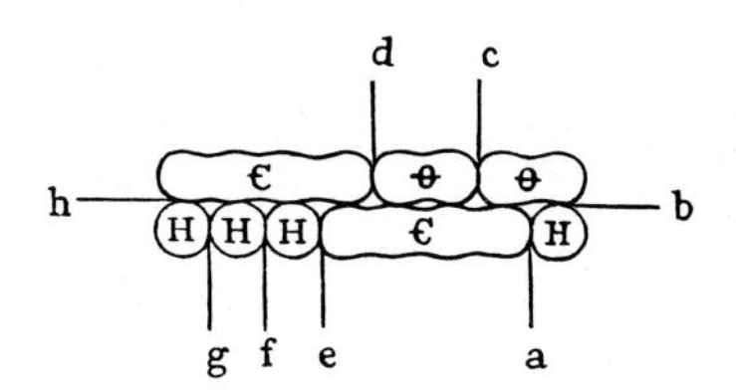

Fig. 3. The chemical structure of acetic acid according to Kekulé[19] (see text).

end of the 18th century[18]. The views of Gerhardt referred to above were followed up by Kekulé who insisted on the pure chemical meaning of the formulas, as emphasized by the example of acetic acid, graphically formulated[19] as shown in Fig. 3. "Fission" at "ab" takes place in salt and ester formation, at "bc" in the formation of thioctic acid; the action of phosphorous chloride causes replacement of O and H at "ac"; formation of acetonitrile involves "ad"; reactions resulting in the formation of methyl compounds involve the position "ed", the action of chlorine or bromine, "hg"[18]. Kekulé, on the other hand, departed from the type theory, which was therefore abandoned*.

"I regard it as necessary and, in the present state of chemical knowledge, possible in many cases, to explain the properties of chemical compounds by going back to the elements themselves which compose these compounds. I no longer regard it as the chief problem of the time to prove the presence of atomic groups which, on the strength of certain properties, may be regarded as radicals and in this way refer compounds to a few types which can hardly have any significance beyond that of mere pattern formulas. On the contrary, I hold that we must extend our investigations to the constitution of the radicals themselves... and from the nature of the elements, deduce both the nature of the radicals and that of their compounds." (ref. 20, translation of Mason[18])

* The theory of types, while forgotten as such, was revived in another field of studies, that of prebiological chemical (abiotic) evolution, as will be shown in a later chapter.

References p. 263

Plate 48. Marcellin Berthelot.

5. The chemistry of natural substances

After the time of extensive collections of proximate principles obtained from organisms, chemists had started to synthesize new substances and to determine their properties. Many examples are at hand, but in the limits of this treatment, we may give as an example the production of derivatives of uric acid by Wöhler and Liebig (see ref. 8, pp. 333–334). The correct formula $C_5N_4H_4O_3$ was found by Mitscherlich. The substance had been discovered by Scheele (1776) in urine and by Prout (1815) in the excrements of birds and serpents. In 1838, Wöhler and Liebig described the following derivatives of uric acid obtained by them by oxidation or by reduction:

allantoin	$C_4N_4H_6O_3$	$C_4N_4H_6O_3$
alloxan	$C_8N_4H_8O_{10}$	$C_4N_2H_2O_4$
alloxanthin	$C_8N_4H_{10}O_{10}$	$C_8N_4H_4O_7+3H_2O$
thionuric acid	$C_8N_6H_{10}O_6S_2$	$C_4N_3H_5SO_6$
uranil	$C_8N_6H_{10}O_6$	$C_4N_5H_5O_3$
dialuric acid	$C_8N_4H_3O_8$	$C_4N_2H_4O_4$
parabanic acid	$C_6N_4H_4O_6$	$C_3N_2H_2O_3$
oxaluric acid	$C_6N_4H_8O_8$	$C_3N_2H_4O_4$
mesoxalic acid	C_3O_4	$H_2C_3O_5$
mycomelic acid	$C_8N_8H_{10}O_5$	$C_4N_4H_4O_2$
uramilic acid	$C_{16}N_{10}H_{20}O_{15}$	
murexid	$C_{12}N_{10}H_{12}O_8$	$C_6N_6H_8O_6$
murexan	$C_6N_4H_8O_5$	$C_4N_5H_5O_3$

(later shown to be the same as muramil)
(second column, formulas determined by Wöhler and Liebig; third column, modern formulas, after Partington[8], p. 333).

When the activity of organic chemists had progressed enough to reach the point where the number of pure organic substances separated from natural materials and produced by synthesis had grown into hundreds of thousands, and where the proximate principles appeared as scattered specimens of the endless possibilities of the synthesizing abilities of the chemist, no basis remained for the necessity of vital force for ruling the organic process of synthesis. Berthelot expressed this new attitude of organic chemists in the following words:

"To ban life from all explanations concerning organic chemistry, this is the objective of our studies. It is only in that perspective that we will succeed in edifying a complete and autonomous science. Chemistry creates its own object. This creative faculty, similar to that of art itself, essentially distinguishes chemistry from natural or historical sciences." (ref. 9, translation by author)

This is not the place to comment on the admirable development of organic

References p. 263

chemistry, or on the innumerable compounds created by the activity of organic chemists; and before putting the question of the rationale of the selected choice of samples characterizing living matter in the midst of the immensity of the innumerable possible compounds, we must know more about the history of the intermediate steps of metabolism and their effectors (Parts III and IV of this History). A knowledge of the organic constituents forming the substance of the cells was of course a requisite before a scientific biochemistry could be founded. This scope has been pursued since the dawn of organic chemistry and is still in progress today.

The field of chemistry of natural substances is covered in a general way in Section II (Vols. 5–11) of this Treatise. Large sections of Lieben's "*Geschichte*"[21] are devoted to the history of our knowledge of natural substances, and more recent developments of this important section of organic chemistry are retraced in the series *Fortschritte der Chemie organischer Naturstoffe*. It is obvious, at least to the present author, that the unravelling of the structure and properties of the organic substances forming the matter of cells is the task of organic chemists. It requires other technical and theoretical qualifications than, for instance, the discovery of the metabolic sequences taking place in cells and the identification of their effectors, or of the unravelling of the molecular basis of physiological and biological concepts, which requires other methods and which can only be accomplished through recourse to experimentation on living organisms. It is the way in which biochemistry is apprehended in this historical presentation. One of its conclusions will nevertheless be that when considering the whole domain of what the ancients called matter-of-life and forces-of-life in the large amount of data gathered by biochemists since the beginning of this century, an increasing number of biochemical topics will, in the following historical phase, move into the domain of the chemist.

One day, which we hope will not be too far away, the fields of biology and pathology will be entirely clarified and become integral parts of the scientific study of molecules, and of their association in the organization of the cell. A history of biochemistry is in fact a history of this progressive shift from the sensorial approach of the naturalist to the molecular approach. But before this development could start, life, in a preliminary methodological stage, had to be banned from organic chemistry.

REFERENCES

1 F. Wöhler, *Poggendorff's Ann.*, 12 (1828) 243.
2 T. Schwann, *Lectures on General Anatomy*, 1841 (Manuscript of his notes and manuscript notes taken by a student), Library of the Department of Biochemistry, University of Liège, Belgium.
3 J. von Liebig, *Die organische Chemie in ihre Anwendung auf Agricultur und Physiologie*, Brunswick, 1840.
4 J. B. Dumas, *Ann. Chim.*, 44 (1830) 273.
5 *Briefwechsel zwischen J. Berzelius und F. Wöhler*, Vol. 1, Leipzig, 1901.
6 D. McKie, *Nature*, 153 (1944) 608.
7 G. J. Goodfield, *The Growth of Scientific Physiology*, London, 1960.
8 J. R. Partington, *A History of Chemistry*, Vol. 4, London, 1964.
9 M. Berthelot, *Chimie Organique Fondée sur la Synthèse*, 2 Vols., Paris, 1860.
10 E. von Meyer, *A History of Chemistry* (translated by G. McGowan) 3rd ed., London, 1906.
11 J. Schiller, *Sudhoffs Arch.*, 51 (1967) 229.
12 F. G. Hopkins, *Biochem. J.*, 22 (1929) 1341.
13 J. H. Brooke, *Ambix*, 15 (1968) 84.
14 T. S. Hall, *Ideas of Life and Matter*, Chicago and London, 1969.
15 E. Kahane, *Bull. Soc. Chim. France*, No. 12 (1963) 4733.
16 C. Gerhardt, *Introduction à l'Étude de la Chimie par le Système Unitaire*, 1848.
17 J. H. van 't Hoff, *Ansichten über die organische Chemie*, Brunswick, 1878–1881.
18 H. S. Mason, *Isis*, 34 (1942–43) 346.
19 F. A. Kekulé, *Lehrbuch der organischen Chemie*, Vol. 1, Erlangen, 1861.
20 F. A. Kekulé, *Ann. Chem.*, 106 (1858) 129.
21 F. Lieben, *Geschichte der physiologischen Chemie*, Leipzig and Vienna, 1935.

Chapter 13

Biocatalysis and the Enzymatic Theory of Metabolism

1. Catalysis

The first catalytic reaction described in the literature is the observation made by Vogel[1] that, at low temperature, oxygen and hydrogen react if charcoal be added to their mixture. Humphrey Davy[2] and his brother Edmund Davy[3] accomplished with platinum the promotion of a number of reactions, a result also obtained by Dulong and Thenard[4,5].

Berzelius proposed to account for these observations by formulating the theory of *catalysis*, and tentatively described the catalytic power as a force apart,

"such that materials may, by their mere presence, and not on account of their chemical affinities, awaken in a substance such affinities as are latent at the temperature in question." (ref. 6, translation of Keilin[7])

Turning to biological phenomena, Berzelius boldly writes:

"We can now assume on very good grounds that in living plants and animals thousands of catalytic processes take place between tissues and fluids, thus giving rise to multitudes of chemical compounds." (ref. 6, translation of Keilin[7])

It must be noted that this formulation of the concept of catalysis was antedated by the discovery of amylase[13] and of pepsin[14]. Mitscherlich[8] accepted the theory of Berzelius and proposed calling the catalysts "contact substances" and the reactions induced by them "decomposition (or combination) by contact."

At the time of the introduction of the catalysis theory by Berzelius, it was already known that a number of reactions were promoted by extracts of living material or by this material itself. Planche[9,10] had demonstrated the oxidation of guaiacum by the roots of plants. Kirchhoff[11] had demonstrated the formation of sugar in an aqueous infusion of flour obtained from germinating grains. Dubrunfaut[12] had obtained the transformation of starch into sugar by a clear, watery extract of barley malt.

References p. 277

Plate 49. Jöns Jacob Berzelius.

2. Identification of enzymes

Payen and Persoz[13] treated malt extract with alcohol and obtained a white amorphous precipitate, soluble in water. The powder was reprecipitated with alcohol from its aqueous solution and it was found to liquefy starch rapidly. To this active substance, Payen and Persoz gave the name of *diastase*⋆. In 1836, Schwann[14] obtained pepsin from pigs' stomach; and in 1837, Wöhler and Liebig[15] extracted emulsin from bitter almonds. In the following years, other ferments were identified: lipase obtained by Claude Bernard[16] from pancreas, invertase from yeast by Berthelot[17], trypsin from pancreas by Kühne[18].

After the recognition of the nature of the agent of alcoholic fermentation as an organism, a difference was made between organized ferments and unorganized ferments, a distinction which persisted until Buchner's observations on yeast juice.

The concept of the existence of intracellular as well as of extracellular enzymes was first accepted by Berthelot[19], Traube[20] and Hoppe-Seyler[21], but the objection was raised by many others that these workers did not isolate enzymes from the cells. There was also a long opposition to the concept of biosynthesis through enzymatic actions. In fact for a long time the concept of enzymatic action was considered to be limited to the preparatory digestive work. It was through the study of alcoholic fermentation that the function of enzymes in cell metabolism was generally accepted after the long controversy which started with the discovery of the living nature of yeast by Cagniard-Latour, by Schwann and by Kützing. The rôle of enzymes in the life of microorganisms was recognized as well as their rôle in the digestive processes of higher forms of life.

3. Opposition to enzyme theories of metabolism

In a review chapter which gives an interesting picture of the concepts on metabolism at the date of its publication, Voit[23] states that the metabolism of the organism is the sum of the metabolism of the cells. On the other hand, he rejects the theory of the participation of "ungeformte

⋆ From διαστασις making a breach, "since it was thought that diastase pierced an imaginary skin on the starch granule and allowed a liquid content to escape." (ref. 22) T. de Saussure[22] had anticipated Payen and Persoz by describing the presence, in malt, of the ferment he called mucine.

References p. 277

Plate 50. Anselme Payen.

Fermente". The idea that there could be a special "ferment" for each kind of molecule appeared to him to be unacceptable.

"Man begreift dabei nicht, warum z.B. manchmal so viel Eiweiss zersetzt und dann weniger Fett angegriffen wird, warum bei Zufuhr von Kohlehydraten das fettzersetzende Ferment nicht tätig ist." (ref. 23)

Resistance persisted to the recognition of the part played by enzymes in the cells of higher forms, until the studies on autolysis revealed the wealth of the enzymatic arsenal of the cells*.

The enzymes have now been recognized as definite chemical species of the nature of proteins. This development took a long time to grow, and a whole century separates the formulation of the enzymatic concept by Berzelius from the first crystallization of an enzyme by Sumner.

Adolph Mayer[26] still considered enzymes as "split organisms", in a characteristic intra-molecular motion which accounts partly for the existence of life. A few years later, Stopes wrote about diastase:

"The name is familiar to any brewer who uses malt, but it is questionable if it represents a real substance, or only an energetic condition of bodies known when inert under other names. The reality of diastatic action is one thing, the substantial existence of a definite compound entitled to that distinctive name is another." (ref. 27)

The first efforts to recognize the nature of enzymes were confused, because they depended on the primitive methods still in use at the time. The current definition of the chemical properties of enzymes included solubility in water and glycerol, and precipitation by alcohol. They were recognized as being non-dialysable. This of course pointed to a protein nature. But the impure state of the preparations obtained led to utter confusion. The tentative determinations of elementary composition were also inconsistent[28].

The views on the mode of action of enzymes were also in a state of great confusion at the end of the 19th century. Two main theories were current at the time. According to one of them, the enzymes were substances able to contract temporary combinations with fermentable substances, and it was possible to regenerate them by splitting these combinations. This theory was accepted by Bunsen and by Hüfner (see Arthus[28]). Another current theory considered that the enzymes acted by molecular vibrations communicated to the substrate. This concept was formulated by the botanist Nägeli[29].

It was also adopted by De Jager[30]. The main idea here was that man had a tendency to require substantiation and that all "forces" of nature had

* For the literature on autolysis the reader is referred to Lieben[24] and to Lambling[25].

References p. 277

Plate 51. Jean François Persoz.

been first considered erroneously as substances (light, electricity, magnetism), and it is of course true that the adepts of the concept of enzymes being definite chemical substances ran the risk of "unjustified substantiation", until Sumner's work proved that they were on the right track. Arthus[28], as well as Nägeli[29] and De Jager[30], claimed that Hüfner's and Bunsen's theory of the protein nature of enzymes was wrong, and he concluded that enzymes have no material existence, and are not definite substances, but "properties of material substances."

4. Enzyme-property

The theory of enzyme-property opposed to enzyme-substance became popular in the dark ages of biocolloidology during which proteins were not recognized as molecules. Duclaux[31] developed the theory of enzyme-property in his book on the chemistry of living matter. In 1933, Duclaux, at the IVth Congrès de Chimie biologique (Paris) still enumerated a series of objections to the concept of enzymes as definite substances[32]. Even seven years after Sumner had crystallized urease, Duclaux clearly paid no heed to the protein nature of enzymes.

Another form of the antisubstantialist views of enzymatic actions was presented in a paper by Herzfeld[33] who proposed considering these actions as resulting from certain mixtures of otherwise common compounds (amino acids and peptides). This idea was hailed by Baur in emphatic terms:

"Der Schleier, hinter dem sich bisher die Natur der Fermente verbarg, ist gelüftet. Es stellt sich heraus, dass Fermente ganz gewöhnliche und zudem wohlbekannte Stoffe sind, deren fein abgestimmte Wirkungen mehr auf Mischungen und auf zusätzlichen Bedingungen beruht, als auf einem Geheimnisvollen chemischen Aufbau." (ref. 34)

Herzfeld had a number of followers: Panzer[35], Von Euler and Svanberg[36] and Biedermann[37]. Baur and Herzfeld[38] even applied their theory to the nature of alcoholic fermentation by yeast juice.

Willstätter[39] and his collaborators tried to reach a conclusion concerning the nature of enzymes by endeavouring to isolate the active substances. From these studies, Willstätter concluded that enzymes consist of colloidal material combined with a chemically active group. The enzymes were obtained in diluted solutions which, although active, gave none of the tests for proteins: it was therefore concluded that enzymes were not of a protein nature.

References p. 277

Plate 52. James B. Sumner.

A member of Abderhalden's school, Fodor[40], introduced the idea according to which enzymes are substances but not defined ones, and are rather substances which, in certain circumstances, behave as enzymes. Fodor typically reflects the attitudes of the colloid chemists:

"Wir können uns allerdings vorstellen, dass im Protoplasma Zellkolloide, wie Proteine, Lipoide, Polysaccharide usw., ein Gallertsystem aufgebaut haben, mit zahllosen Grenzflächen und kapillaren Räumen, die durch Elektrolyte bzw. ihre Ionen absorbtiv beladen sind, wodurch diese zustandsbestimmend und -erhaltend wirken. Auf solchen Grenzflächen betätigen sich elektrische Kräfte und fermentaktive Stoffe bzw. Gruppen. Mit der Zerstörung der Protoplasmastruktur werden jene nicht total vernichtet, diese teilweise konserviert und von ihren zugehörigen Fermentkolloiden (die feinen Grenzflächen sind ja nicht mehr vorhanden) mehr oder weniger grober Dispersion festgehalten oder "getragen"." (ref. 40)

Since 1897, the date of the preparation of yeast juice by Buchner, the progress achieved in the study of alcoholic fermentation brought in a number of observations in favour of the role of enzymes in the metabolism of yeast cells, and consequently in other cells. This we will consider in later chapters. In spite of this, as late as 1927 (after the crystallization of urease by Sumner), Kostytschew *et al.*[41] again tried to explain away the results of Buchner and his followers.

5. Enzymes as proteins

The biochemists' world was deeply shaken by the crystallization by Sumner[42] of urease from jack bean. Willstätter, as well as other chemists, dismissed the achievement as being merely the crystallization of a protein "carrier" of the enzyme.

The following years saw the development of Northrop's work. He purified and crystallized several digestive enzymes, all the enzymes crystallized proving to be proteins[43]. The controversy that had been raging for a hundred years on the nature of enzymes was thus ended.

"The early work on enzyme purification met with a good deal of criticism on the grounds that it was "unphysiological" to separate enzymes from cells and that only work on undamaged cells was valuable. Later, when the first protein crystals of high enzyme activity had been obtained, there was considerable skepticism towards the claim that the enzymes themselves had been crystallized; it was suggested that the enzyme was present as a mere impurity, adsorbed on crystals of an inert protein. Northrop, however, produced conclusive evidence that his crystals were crystals of the enzymes themselves, and all the enzymes which have since been isolated in the pure state have proved to be proteins. Thus the conception of an enzyme has undergone a development, first from a vague influence or property in certain preparations to definite chemical substances, and finally to specific proteins." (Dixon and Webb[44])

The purification of intracellular enzymes became active around 1937, and

References p. 277

Plate 53. John H. Northrop.

as we shall see, the number of known enzymes has greatly increased since.

The kinetics of enzyme actions have been the subject of a large number of studies. In short, it may be said that the pioneer work in this field was accomplished on saccharose by Victor Henri in Paris and by Brown in Birmingham in the beginning of our century. These workers concluded that enzyme and substrate form a compound which breaks down at a certain rate into enzyme and product. In an ordinary reaction, in accordance with the law of mass action, the rate is proportional to the concentrations of the reactants but, as acutely observed by Henri as well as by Brown, in the enzyme reaction the rate increases as the concentration of reactants increases, to a limiting value corresponding to a concentration of reactants at which the enzyme is saturated and does not increase further with new additions of reactants. Henri worked out a mathematical theory which was revived and developed by Michaelis in 1913. Enzyme kinetics has since given rise to a number of complicated expressions. (On the history of the physical chemistry of enzyme kinetics, the reader is referred to Vol. 12 of this Treatise, as well as to a detailed study by Segal[45].)

Bertrand[46] introduced the concept of coenzyme. He suggested that an enzyme comprises two constituents, one active and the other inactive by itself but enhancing the action of the first. For instance, manganese exerts a catalytic effect on the oxidation of the diphenols, but the addition of a certain protein, called coenzyme by Bertrand, enhances the action. As remarked by Dixon[47]:

"It is interesting that coenzyme functions are ascribed to the enzyme and what we should now call the inorganic activator or cofactor was regarded as the catalyst. There were many other confusions before the true mode of action of enzymes was cleared up. For example in 1902 Schardinger found in milk an enzyme which we now know catalyses the oxidation of aldehyde by the oxidizing dye methylene blue. But he regarded the aldehyde as the catalyst, enabling the *enzyme* to reduce the dye." (Dixon[47])

Coenzymes were later recognized as providing the functional link between one enzyme and another in the biochemical systems of metabolic pathways[47].

The recognition of enzymes as well-defined proteins was linked to the recognition of proteins as well-defined macromolecules (see Chapter 15), a concept which put an end to the sterile interlude of biocolloidology inaugurated in 1861 and which lasted until well into the forties (see Chapter 14).

Before the enzymes were recognized as specific proteins, an analogy with such catalysts as platinum black had led to the idea that enzymes were surface catalysts, a concept that was superseded by the notion of the enzymes

References p. 277

as definite molecules entering into the reaction. That the shape of the substrate molecule as well as its structure is important has been clearly proven since, confirming the suggestion made by Fischer in 1894 that the enzyme and the substrate fit together like a key and a lock.

The recognition of the specificity of enzyme actions and of their detailed structure has led to the concept of an "active centre", a special configuration on the protein macromolecule of an enzyme, responsible for the interaction between enzyme and substrate. The active centre may be a sequence of amino acid residues, or it may contain such substances as flavin, pyridoxal or heme, for instance. The shape of the active centre or of another part of an enzyme fits, on the substrate molecule, a signal which is itself an aspect of its configuration, or it fits the shape of the substrate molecule. (On the historical aspects of the concept of the active centre, the reader is referred to Vols. 12–16 of this Treatise.)

Since the concept of enzymes as proteins of a definite nature became standard, a large number of enzymes have been discovered, as we shall record in the chapters on the history of the unravelling of metabolic pathways (Parts III and IV). Webb[48] estimates that they will eventually reach the number of between two to three thousand. As the number of known enzymes increased, it became clear that in spite of the large number of different reactions catalysed by enzymes, all these reactions belong to a limited number of reaction types. This enabled the *Commission on Enzymes* of the International Union of Biochemistry to produce a scheme for the classification and nomenclature of enzymes*.

---

* See Vol. 13 (2nd ed.) of this Treatise.

REFERENCES

1 F. C. Vogel, *J. Chem. Phys*, 4 (1812) 142.
2 H. Davy, *Philosoph. Trans.*, (1817) 77.
3 E. Davy, *Philosoph. Trans.*, (1820) 108.
4 P. L. Dulong and L. J. Thenard, *Ann. Chim. (Phys.)*, 23 (1823) 440.
5 P. L. Dulong and L. J. Thenard, *Ann. Chim. (Phys.)*, 23 (1823) 380.
6 J. J. Berzelius, *Lehrbuch der Chemie*, Vol. VI, Leipzig, 1837.
7 D. Keilin, *The History of Cell Respiration and Cytochrome* (prepared for publication by J. Keilin), Cambridge, 1966.
8 E. Mitscherlich, *Ann. Chim. (Phys.)*, 3e Sér., 7 (1843) 15.
9 L. A. Planche, *Bull. Pharm.*, 2 (1810) 578.
10 L. A. Planche, *J. Pharm. Chim. (Paris)*, 6 (1820) 16.
11 G. S. C. Kirchhoff, *J. Pharm. Chim. (Paris)*, 2 (1816) 250.
12 A.-P. Dubrunfaut, *J. Tech. Ökon. Chem.*, 9 (1830) 156, 157.
13 A. Payen and J. F. Persoz, *Ann. Chim. (Phys.)*, 53 (1833) 73.
14 T. Schwann, *Arch. Anat. Physiol. (Leipzig)*, (1836) 90.
15 F. Wöhler and J. von Liebig, *Ann. Phys. (Leipzig)*, 41 (1837) 345.
16 Cl. Bernard, *J. Pharm. Chim. (Paris)*, 15 (1849) 336.
17 M. Berthelot, *Compt. Rend.*, 50 (1860) 980.
18 W. Kühne, *Verhandl. Naturh.-Med. Ver. (Heidelberg)*, (1877) 190.
19 M. Berthelot, *Compt. Rend.*, 51 (1856) 980.
20 M. Traube, *Gesammelte Abhandlungen*, Berlin, 1899.
21 F. Hoppe-Seyler, *Arch. Ges. Physiol.*, 12 (1876) 1.
22 J. R. Partington, *A History of Chemistry*, Vol. IV, London, 1964.
23 C. von Voit, in *Hermann's Handbuch der Physiologie*, Vol. 6, Part I, Leipzig, 1881.
24 F. Lieben, *Geschichte der physiologischen Chemie*, Leipzig and Vienna, 1935.
25 E. Lambling, *Précis de Biochimie*, Paris, 1919.
26 A. Mayer, *Die Lehre von den chemischen Fermenten*, Heidelberg, 1882.
27 H. Stopes, *Malt and Malting*, London, 1885.
28 M. Arthus, *Nature des Enzymes*, Paris, 1896.
29 C. von Nägeli, *Theorie der Gärung*, Munich, 1879.
30 L. de Jager, *Virchow's Arch.*, 121 (1890) 182.
31 J. Duclaux, *La Chimie de la Matière Vivante*, Paris, 1910.
32 J. Duclaux, Quelques Aspects Physico-Chimiques du Problème des Diastases, *4th Congr. Chim. Biol.*, Paris, 1933.
33 E. Herzfeld, *Biochem. Z.*, 68 (1915) 402.
34 E. Baur, *Jahrb. Chem.*, 25 (1915) 384.
35 T. Panzer, *Z. Physiol. Chem.*, 93 (1915) 316.
36 H. von Euler and O. Svanberg, *Z. Physiol. Chem.*, 110 (1920) 175.
37 W. Biedermann, *Fermentforschung*, 2 (1919) 458.
38 E. Baur and E. Herzfeld, *Biochem. Z.*, 117 (1921) 96.
39 R. Willstätter, *Untersuchungen über Enzyme*, Vol. 2, Berlin, 1928.
40 A. Fodor, *Das Fermentproblem*, Dresden, 1922.
41 S. Kostytschew, G. Medwedeid and H. Kardo-Sysojewa, *Z. Physiol., Chem.*, 168 (1927) 244.
42 J. B. Sumner, *J. Biol. Chem.*, 69 (1926) 435.
43 J. H. Northrop, M. Kunitz and R. M. Herrott, *Crystalline Enzymes*, New York, 1948.
44 M. Dixon and E. C. Webb, *Enzymes*, London, 1958.

45 H. L. Segal, in P. D. Boyer, H. Lardy and K. Myrback (Eds.), *The Enzymes*, 2nd ed., Vol. 1, New York, 1959.
46 G. Bertrand, *Compt. Rend.*, 124 (1897) 1032.
47 M. Dixon, in J. Needham (Ed.), *The Chemistry of Life*, Cambridge, 1970.
48 E. C. Webb, *ICSU Rev.*, 5 (1963) 174.

Chapter 14

The Dark Age of Biocolloidology

The concept at the basis of the biochemical theory of colloids had been formulated in 1861 by Graham who states about colloids that

"Their peculiar physical aggregation with the chemical indifference referred to appears to be required in substances that can intervene in the organic process of life. The plastic elements of the animal body are found in this class. As gelatine appears to be its type, it is proposed to designate substances of the class as *colloidal* and to speak of their peculiar form of aggregation as the *colloidal condition of matter*. Opposed to the colloidal is the crystalline condition. Substances affecting the latter form will be classed as *crystalloids*. The discussion is no doubt one of intimate molecular constitution." (ref. 1)

Graham classified as colloids substances unable to penetrate certain membranes, that diffuse slowly and that cannot be crystallized. The first two characters apply to the majority of proteins and the third to certain proteins. But this did not disturb Graham who states that no sharp discontinuity exists between colloids and crystalloids:

"A departure from its normal condition appears to be presented by a colloid holding so high a place in its class as albumen. In the so-called blood-crystals of Funke, a soft and gelatinous albuminoid body is seen to assume a crystalline contour. Can any fact more strikingly illustrate the maxim that in nature there are no abrupt transitions, and that distinctions of class are never absolute?" (ref. 1)

A typical example of overvalorization is found in the following sentence of Graham:

"The colloidal is, in fact, the dynamical state of matter, crystalloidal being the statical condition. The colloid possess *ENERGIA*. It may be looked upon as the probable primary source of the force appearing in the phenomena of vitality." (ref. 1)

Proteins were from then on considered as colloids. Later, Ostwald[2] developed the theory of the colloidal state of matter, and substances that have nothing in common with proteins were defined as colloids, such as gold salts, soap solutions, colloidal solutions of tannic acid, etc. No distinction was made

References p. 284

between what we now call macromolecules made up of atoms held together by covalent bonds, and colloidal particles composed of any molecules of ordinary size held together by intermolecular "secondary valences".

It took some time after the publications of Graham before the concept of biocolloids became generally widespread. There was a considerable resistance to its diffusion. In 1897, for instance, Neumeister, in his *Lehrbuch der physiologischen Chemie* only casually mentions it and rejects it:

"... die verbreitete Annahme, dass die so genannten Kolloïde sich nicht in wirklicher Lösung befänden, ist durchaus willkürlich. Die Ursache warum gewisse Stoffe diffundieren, andere nicht, ist lediglich darin zu suchen, dass die Moleküle der nicht diffusiblen substanzen wegen ihre bedeutenden Grösse die feinen Poren der Membranen nicht passieren können." (ref. 3)

But increasing overvalorization was apparent in the publications concerning biocolloidology. It is in this overvalorization that we find the element of seduction which the doctrine exerted on biologists. Its supporters claimed that many biological phenomena such as parthenogenesis, muscle contraction, production of action currents in nerves, heart activity, ciliary movements, etc. were influenced by inorganic ions according to a series of degrees of influence parallel to the influences of the same series of ions on heat coagulation, on lecithin precipitation, etc. The induction that biological phenomena were the results of changes introduced by ions in agglutination, lytic processes, dispersion, hydration or dehydration of colloidal micelles believed to compose the protoplasmic "gel" became very widespread. The appeal of this "micellar biology" was similar to that exerted in more recent years by "molecular biology", as shown by the tremendous success of the concept formulated by Gortner when he wrote in the preface of his *Outlines of Biochemistry:*

"All of the reactions and interactions which we call life take place in a colloidal system, and the author believes that much of the "vital energy" can in the last analysis be traced back to energies characteristic of surface films and interfaces." (ref. 4)

Innumerable statements could be found in the literature, showing an incredible amount of overvalorization of the concepts of biocolloidology, called upon to explain all the mysteries of life. A few of them, taken from a number of scientific languages, are typical of this situation and will dispense with citing an enormous amount of now useless literature.

"Graham a créé le mot *colloïdal* pour caractériser l'état physique des corps analogues à la colle de gélatine. D'après lui (et nous partageons absolument cette opinion), la forme colloïdale serait une sorte *d'état transitoire instable ou dynamique, dont l'état statique* est la *forme cristal-*

lisée. En fait les matières albuminoïdes sont aptes à se transformer sous les moindres influences, action de la chaleur et du froid, des gaz, des sels neutres, des ferments, etc. Leur simple dilution dans l'eau, l'agitation ou le repos, le passage à travers certaines membranes inertes, etc. peuvent leur imprimer des changements importants tels que la solubilité ou l'insolubilité.

De nature chimique indifférente, faiblement unis dans l'économie à une grande masse d'eau, ces colloïdes fluides ont une mollesse qui les rend propres, comme l'eau elle-même, mais moins brutalement qu'elle, aux phénomènes de *diffusion*; ils sont lentement pénétrables aux réactifs, et leurs molécules servent d'intermédiaires et comme d'amortisseurs aux plus délicates actions physico-chimiques. Passant difficilement à travers les membranes, même si celles-ci sont, comme le papier à filtrer, percées de pores sensibles, les matières albumineuses se conservent dans les cellules sans diffuser au dehors. Cette difficile diffusion, la lourdeur des molécules de ces corps, leur faible conductibilité pour la chaleur et l'électricité, leur indifférence chimique, concourent à ralentir les réactions qui se produisent dans nos tissus et nos humeurs. Grâce à ces propriétés, les réactions que la vie met en jeu se poursuivent sans secousses, successivement, assurant ainsi au fonctionnement des organes une production d'énergie progressive issue de ces transformations lentes mais continues." (Gautier[5])

"Die lebende Substanz, d.h. diejenige, auf welcher sich die Lebenserscheinungen abspielen, zeigt, wie gesagt, die Eigenschaften des halbflüssigen Aggregatzustandes. Drei Umstände sind besonders zu beachten. Erstens: halbflüssige Gebilde sind durch Zug- und Druckkräfte leicht deformierbar gleichzeitig ob diese Kräfte von aussen auf sie einwirken oder die Energie innerhalb der Masse selbst auftritt. Auf diese Weise kommen die für die Lebenserscheinungen charakteristischen Eigenbewegungen der lebendigen Substanz zustande..." (Rosenthal[6])

"... un grand nombre de phénomènes biologiques observés ne trouvent d'interprétation satisfaisante que si on les considère comme les conséquences de formations colloïdales et logiquement il doit en être ainsi car la matière vivante est un complexe dans lequel prédomine l'état colloïde." (Rocasolano[7])

"Les milieux humoraux, les protoplasmes cellulaires, les tissus des animaux et des végétaux au sein desquels s'effectuent les processus vitaux, sont tous des colloïdes. Toutes les transformations, toutes les réactions qui conditionnent la croissance et la nutrition des êtres vivants portent sur des éléments d'une nature particulière, sur des substances colloïdales.... Ce qui caractérise avant tout un colloïde c'est donc sa structure complexe particulière *constamment variable*, avec tendance à la soudure, au grossissement des éléments granulaires qui le composent et à la floculation...". (Lumière[8])

At the time of the foundation of the *Kolloid-Gesellschaft* in 1922, Abderhalden, speaking at the foundation meeting in Leipzig, ended his speech by saying "Wir Biologen begrüssen die Gründung der "Kolloid-chemischen Gesellschaft" auf der Wärmste."[9]

Mathews, in his *Physiological Chemistry*, referring to the specializations of different parts of cells, which were later to be shown to depend on different organelles, attributes them to biocolloids.

"It is by means of the colloids of a protein, lipoid or carbohydrate nature, which make up the substratum of the cell that this localization of chemical reactions is produced; the colloids furnish the basis for the organization or machinery of the cell; and in their absence there could be nothing more than a homogenous conglomeration of reactions... The colloids localize

References p. 284

the cell reactions and furnish the physical basis of its physiology; they form the cell machinery." (Mathews[10])

In P. Thomas's *Cours de Chimie Biologique* mechanical aspects of the organism are attributed to the colloidal state, and he attributes to the colloidal state the rapid mobilization of the components of organisms.

"Les colloïdes combinent les qualités de l'état solide avec celles des liquides par leur développement superficiel. Si on pense à l'énergie développée en quelques heures dans une ascension en montagne par un alpiniste; à la consommation énorme de protéines qui a lieu chez un fébricitant; aux quantités de substances mobilisées par un arbre pendant les quelques jours où il se couvre de feuilles, etc. on sent bien que des corps non colloïdaux, avec la faible surface dont ils disposent, ne pourraient assurer une aussi rapide mobilisation des substances." (Bechhold[13]) (ref. 11)

Rogers emphasizes the biological importance of the relations between colloids and crystalloids:

"Undoubtedly, the colloids form the stable substrate of the living substance. Associated with the colloidal elements are certain crystalloid substances which penetrate to all parts of the colloidal complex and which here and there become temporarily bound up in the complex. They are then freed to become again, at some other point, a part of the living machine. Some of the phenomena which are called *vital* are beyond question the direct result of the shifting here and there of molecules of one sort or another in the colloidal complex." (Rogers[14])

In 1928, Heilbrunn published a whole book devoted to the cell "protoplasm", in which he refers in passing to mitochondria in the following terms:

"... but at best the mitochondria concept is not a very precise one." (Heilbrunn[15])

A year later Vlès[16] stated that according to physiological conditions "protoplasm" could shift from the state of a "gel" to that of a "sol". This lability is considered by Kiessel[17] to be one of the conditions of life.

"Die Zustand der chemischen und physikalischen Labilität des Protoplasmas ist sicher eine Vorbedingung der Lebenserscheinungen und die hohe Reaktionsfähigkeit des Protoplasmas scheint ein Zeugnis fur diese Labilität abzulegen." (Kiessel[17])

"There is no more important field in colloidal chemistry than the field which deals with the water relations of the biocolloids, and there is abundant evidence that these relationships determine in a large measure the vital activities of organisms. (Gortner[18])

"Da nun in den Geweben von Zellwand zu Zellwand die verschiedensten kolloidal Lösungen, wenn auch nur in winzigen Dimensionen, aneinandergrenzen, so geht daraus schon die Wichtigkeit des kolloidalen Zustandes und der Grenzflächenphasen hervor: sozusagen alle Reaktionen in der lebenden Materie sind Grenzflächenreaktionen.

"Die Besonderheit einer Grenzfläche liegt in erster Linie darin, dass (*1*) ihre Bausteine, falls es sich um Kolloide handelt, in ihr besonders angereichert sind und dass (*2*) die Bindungskräfte dieser Bausteine, auch wenn es sich nicht um kolloidale Körper handelt, nicht derartig von allen Seiten beansprucht werden, wie dies im Innern der begrenzten Phase, z.B. der Zelle, der Fall ist. Das hat zur Folge, dass die Grenzfläche im allgemeinen durch eine hohe Reaktionsbereitschaft ausgezeichnet ist. Vorgänge und Reaktionen verlaufen in den Grenzflächen meist erheblich rascher als in den von ihnen umschlossenen Phasen. In dieser erhöhten Reaktionsbereitschaft und in dem raschen Ablauf der Reaktionen liegt auch einer der wesentlichen Gründe dafür, warum in der lebenden Zelle—jeder Zellvorgang ist wie gesagt ein Grenzflächenvorgang—Reaktionen bei Körpertemperaturen verlaufen, welche sonst ausserhalb des Grenzflächensystems der Zellwände erst bei wesentlich höheren Temperaturen stattfinden würden." (Brandt[19])

"All the chemical reactions associated with life take place between substances in aqueous solutions and aqueous colloidal systems, and the properties of these solutions and systems largely determine tissue organisation and function." (Fearon[20])

As these quotations indicate, the colloids were considered as explaining the nature of life. When it became clear in the twenties that no place existed in biochemistry for the concept of micelles, supposed to exist in all kinds of states of aggregation, whose evolution ruled the course of life phenomena, it is surprising to realize that the concept of biocolloidology based mainly on studies of degraded or impure preparations remained accepted in certain circles until well into the forties. In the textbook of Gortner, *Outlines of Biochemistry*[4], the whole presentation is still based on the colloidal theory, and in its subject index the words "colloids", "sols" and "gels" appear in almost a hundred entries, whereas in any reputable textbook of biochemistry, published after World War II, these words are not even mentioned*.

* This does not mean that a number of textbooks do not repeat, for reasons unknown, a stereotyped treatment in which proteins appear as dispersed and not dissolved. Cell organelles are still also sometimes defined as "colloidal droplets".

References p. 284

REFERENCES

1 T. Graham, *Philosoph. Trans. Roy. Soc. (London)*, (1861) 183.
2 W. Ostwald, *Kolloid-Z.*, 1 (1907) 331.
3 R. Neumeister, *Lehrbuch der physiologischen Chemie*, Jena, 1897.
4 R. A. Gortner, *Outlines of Biochemistry*, 2nd ed., New York, 1938.
5 A. Gautier, *Leçons de Chimie Biologique*, 2nd ed., Paris, 1897.
6 J. Rosenthal, *Lehrbuch der allgemeinen Physiologie*, Leipzig, 1901.
7 A. de Gregorio Rocasolano, *Éléments de Chimie Physique Colloïdale*, Paris, 1920.
8 A. Lumière, *Rôle des Colloïdes chez les Êtres Vivants, Essai de Biocolloïdologie*, Paris, 1921.
9 E. Abderhalden, in *Kolloidchemie der Gegenwart*, (special issue), *Kolloid-Z* 31 (1922) Heft 5.
10 A. P. Mathews, *Physiological Chemistry*, 3rd ed., London, 1924,
11 P. Thomas, *Cours de Chimie Biologique*, Vol. 1, Paris, 1926.
12 J. Duclaux, *Les Colloïdes*, Paris, 1920.
13 H. Bechhold, *Die Kolloide in Biologie und Medizin*, 4e Aufl., Dresden and Leipzig, 1938.
14 C. G. Rogers, *Textbook of Comparative Physiology*, 3rd ed., New York and London, 1938.
15 L. V. Heilbrunn, *The Colloid Chemistry of Protoplasm*, Berlin, 1928.
16 F. Vlès, *Précis de Chimie Physique*, Paris, 1929.
17 A. Kiessel, *Chemie des Protoplasmas*, Berlin, 1930.
18 R. A. Gortner, *Selected Topics in Colloid Chemistry with Especial Reference to Biochemical Problems*, Ithaca, N.Y., 1937.
19 W. Brandt, *Physiologische Chemie für Medizinier und Biologen*, Stuttgart, 1939.
20 W. R. Fearon, *An Introduction to Biochemistry*, 3rd ed., London, 1946.

Chapter 15

Recognition of the Proteins as Truly Defined Macromolecules

The field of protein studies endured longer than the other chapters of the chemistry of natural substances, a domain not fully transferred to a section of organic chemistry, an aspect which, though outside the domain of our history, has retarded the development of scientific biochemistry and consequently is of interest in the present context.

We have related how biologists of the 19th century considered, as did Robin and Verdeil, that proteins were not defined chemical entities. The organic chemists long remained influenced by Liebig's[1] concept, according

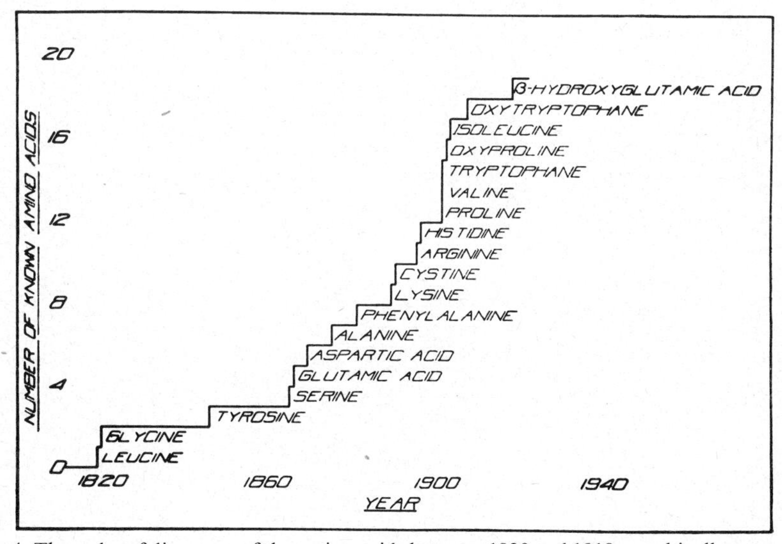

Fig. 4. The order of discovery of the amino acids between 1820 and 1918, graphically represented by plotting the number of amino acids that have been isolated from a protein source as ordinates, and the year of their discovery as abscissa. (From Cohn[3])

References p. 294

Plate 54. Hermann Staudinger.

to which all proteins (fibrin, casein, albumin, etc.) were identical in composition, a concept set aside when Dumas and Cahours[2], by means of their newly developed method for the determination of nitrogen, showed differences in the elementary composition of proteins.

By different techniques of hydrolysis, the building blocks of the proteins, the amino acids, were progressively identified. Fig. 4 borrowed from Cohn[3] reports the dates of the isolation of amino acids from protein hydrolysates between 1820 and 1918.

How the amino acids are associated in the protein structure was the subject of a large number of studies, leading to diverse theories, the history of which has been retraced by Vickery and Osborne[4].

Fischer[5] and Hofmeister[6] clarified the situation by the formulation of the peptide hypothesis, and in spite of many concurrent theories (see Vickery and Osborne[4]), the peptide theory has become a paradigm. Organic chemists were in the meantime progressing in their studies of protein structure which finally led to the present-day recognition of proteins as well-defined macromolecules composed of peptide chains with unique sequences that can be precisely defined. (For a history of this development, the reader is referred to Vols. 7 and 8 of this Treatise.)

In 1901, Schulz[7] published a book on the crystallization of proteins showing that the crystals of many of them were easy to prepare and that their character as genuine crystals had been established by protein chemists as well as by crystallographers. Another book by Schulz, published[8] in 1903, portrays the problem of the molecular weight of proteins. Although he considers the situation as unsatisfactory, there is no doubt that Schulz leans towards the recognition of proteins as true large molecules. But the whole current of biocolloidology (Chapter 14) upheld an entirely different viewpoint and considered that the "solutions" of colloids were in fact suspensions and that the physico-chemical laws of solutions were inapplicable to them.

The rejection of the concept of a single phase in the case of "solutions" of proteins or other "colloids" was strongly supported, in particular by W. O. Ostwald in Germany and by W. Bancroft in the United States.

The macromolecular nature of many natural or synthetic substances was established by Staudinger[9]. The idea that "high" polymers "are composed of covalent structures many times greater in extent than those occurring in simple compounds, and that this feature alone accounts for the characteristic properties which set them apart from other forms of matter"[10]

References p. 294

Plate 55. S.P.L. Sørensen.

goes back, in the case of carbohydrates and proteins, to Hlasiwetz and Habermann[11]. But Staudinger provided the arguments which, within a couple of decades, were to bring the concept of macromolecule to the state of a paradigm*. (For the historical aspects of this development, see Staudinger[12], 1947.)

A vivid sketch of the history of the macromolecular concept is drawn by Olby, as follows:

"Staudinger's study of polymers began in 1920 when he was at the Swiss Federal Technical University in Zurich. That year he delivered a spirited attack on the recent assumption of secondary forces of association in polymers, himself favoring primary Kekulé bonds throughout. Two years later he used the term macromolecule, and in 1934 he defined it as those particles in which the individual atoms are bound by normal valency forces. When he left Zurich in 1925 to take the chair of chemistry at the German university of Freiburg im Breisgau he devoted his lecture at the Zurich Chemical Society to his evidence in favor of the macromolecule concept. This formed the subject of his address in Düsseldorf a year later.

"On both occasions he met with a hostile reception. The Zurich meeting finished in uproar and Staudinger concluded the stormy discussion defiant but not aggressive, quoting the words of Luther, "Here I stand, I cannot do otherwise". Let me quote some of the advice he received.

"Dear Colleague, leave the concept of large molecules well alone; organic molecules with a molecular weight above 5000 do not exist. Purify your products, such as rubber, then they will crystallise and prove to be lower molecular substances."

"Organic molecules with more than 40 carbon atoms do not exist."

"Molecules cannot be larger than the crystallographic unit cell, so there can be no such thing as a macromolecule."

"There were three reasons for this violent opposition to Staudinger's macromolecule concept. The aggregate theory we have seen harmonized with the concept of colloidal properties as being supramolecular in origin, which was the orthodox view at that time. The new science of X-ray crystallography, which had only just been applied successfully to cellulose by Herzog and his colleagues at the newly founded Kaiser Wilhelm Institute for fiber chemistry, was mistakenly thought to support the concept of small polymer molecules. A molecule bigger than the crystallographic unit cell cannot exist. Third, Staudinger's most important evidence—chain length of polymers—did not rest on a theoretical relationship derived from first principles but on his empirical Staudinger Rule for specific viscosity. Had he been dealing with spherical molecules which in solution obey the Einstein–Smoluchowski equation, then he could have derived molecular weights, or rather particle weights, in a logical manner from viscosity measurements**, but precisely because he was dealing with long chain molecules his viscosity was of the non-Newtonian type, and he was forced to extrapolate to small polymers accessible to traditionally acceptable methods of molecular weight estimation in order to calibrate his readings. It is interesting to note here that Staudinger had arrived at the relationship for specific viscosity in a fit of desperation brought about by the refusal of the authorities to supply him with the funds necessary to purchase an ultracentrifuge (Dr. M. Staudinger, personal communication).

* The biological importance of the concept was clearly understood by Staudinger, as will be pointed out in the Chapter on the history of molecular biology, in Part V of this History.

** Even with spherical particles there is the problem of solvation, which causes a deviation from theory.

References p. 294

Plate 56. Edwin J. Cohn.

"His other line of evidence was that methylation, acetylation, bromination, and hydrogenation of polymers does not alter their molecular weight substantially. How, he argued, could they exhibit this identity if they are held together by secondary forces which must be destroyed before the chemical reaction can take place? And had not the aggregate theory upholders attributed the secondary forces to residual affinities which are used up in methylation and hydrogenation? How then could rubber, an unsaturated hydrocarbon, and hydrogenated rubber, a saturated one, both have secondary forces in them? Resistance continued, and it was not until 1953, 33 years after he had introduced the term "macromolecule" that the 72-year-old Staudinger received the Nobel Prize for chemistry." (ref. 13)

A few other early investigators were led by their observations to consider that cellulose or starch were composed of very large true molecules. But the theory of the "colloidal state" with its overvalorized "biological" virtues was very popular among biologists. On the other hand, all the aspects of physically associated substances attracted great interest. Last but not least, the organic chemists were reluctant to consider the existence of such huge molecules.

In regard to proteins, the study of Sørensen[14] on ovalbumin, published in 1917, exerted a deep influence after the end of the first world war. Not only had he shown that the solubility of ovalbumin in ammonium sulphate solutions could be defined by the phase rule of Gibbs but he also measured very accurately the osmotic pressure of ovalbumin and calculated a molecular weight of 34000, an estimate near the true value. The consequences of the work of Sørensen were not easily accepted by the colloid chemists. The present author, who worked with Edwin Cohn in the twenties, has heard him tell the story related by John Edsall, who also worked in that laboratory at the time:

"At a scientific meeting shortly after 1920, Cohn had an exchange of remarks with Wilder D. Bancroft, the well-known colloid chemist, which went somewhat as follows:

Cohn: Sørensen has measured the osmotic pressure of egg albumin, and finds a molecular weight of 34000.

Bancroft: Yes, yes, I understand. He is measuring a system of molecular aggregates. That is the molecular weight of the aggregate.

Cohn: But the tryptophan content and sulphur content of ovalbumin have also been determined, and they give a *minimum* molecular weight of 34000.

Bancroft: Ah! Then in that case I would say that the aggregation factor is unity!" (ref. 15)

Cohn was a student, not only of Osborne and of Sørensen, but also of Jacques Loeb whose contributions were also critical in arriving at the true concept of the nature of proteins[16].

Around 1924, doubts were cast on the theory of the macromolecular nature of proteins by a view according to which the proteins, when dissolved

References p. 294

Plate 57. The Svedberg.

in phenol, fall apart into units with molecular weights of only a few hundred. The conclusion was that the proteins were aggregates of small molecules held together by secondary valences[17,18]. Cohn and Conant[19] showed, by measurement of freezing points of protein solutions in phenol, that the values obtained were the result of the presence of water as an impurity in the phenol. When the water content of the phenol was fixed by equilibrating it with calcium chloride, the addition of proteins produced negligibly small changes in freezing point depressions. The whole evidence showing that proteins were macromolecules and true electrolytes was summarized by Cohn[3] in 1925. In the same year appeared the studies of Adair[20] on the osmotic pressure of haemoglobin, showing that it was four times bigger than had been believed.

When Svedberg started his work with his ultracentrifuge in the belief that proteins were heterogeneous collections of micelles of different sizes, he was startled to observe that the sedimentation of haemoglobin revealed that this protein was of a very regular build[21]. The use as criteria of protein purity, of the ultracentrifuge, of electrophoresis introduced by Tiselius[22,23], as well as of the solubility test, were of great importance in resolving protein mixtures into their macromolecular components.

As stressed by Edsall in a thoughtful essay[15] to which the present chapter is greatly indebted, the application of these new techniques revealed the heterogeneity of many protein preparations considered as pure, and this in some cases shook the faith in the existence of pure homogeneous protein preparations to such an extent that a number of scientists developed the attitude of considering all protein preparations as "microheterogeneous"[24], *i.e.* as heterogeneous collections of similar but not identical molecules.

This theory was disposed of by the development, started by Sanger[25], of the techniques of establishing the sequences of amino acids in proteins. This completely established that proteins are well-defined molecules composed of peptide chains with unique sequences. The number of these known sequences is increasing every year, as testified by the increasing size of the yearly published *Atlas of Protein Sequences and Structure*[26].

References p. 294

REFERENCES

1 J. von Liebig, *Ann.*, 39 (1841) 129.
2 J. B. Dumas and A. Cahours, *Ann. Chim. Phys.*, 6 (1842) 385.
3 E. J. Cohn, *Physiol. Rev.*, 5 (1925) 349.
4 H. B. Vickery and T. B. Osborne, *Physiol. Rev.*, 8 (1928) 393.
5 E. Fischer, *Untersuchungen über Aminosäuren, Polypeptide und Proteine*, Berlin, 1906.
6 F. Hofmeister, *Ergeb. Physiol.*, 1 (1902) 759.
7 F. N. Schulz, *Die Krystallisation von Eiweisstoffen und ihre Bedeutung für die Eiweisschemie*, Jena, 1901.
8 F. N. Schulz, *Die Grösse des Eiweissmoleküls*, Jena 1903.
9 H. Staudinger, *Ber.*, 53 (1920) 1073.
10 P. J. Flory, *Principles of Polymer Chemistry*, Ithaca, N.Y., 1953.
11 H. Hlasiewetz and J. Habermann, *Ann. Chem. Pharm.*, 159 (1871) 304.
12 H. Staudinger, *Makromolekulare Chemie und Biologie*, Basel, 1947.
13 R. Olby, *J. Chem. Educ.*, (March 1970) 168.
14 S. P. L. Sørensen, *Compt. Rend. Trav. Lab. Carlsberg*, 12 (1917) 1.
15 J. T. Edsall, *Arch. Biochem. Biophys.*, Suppl. 1 (1962) 12.
16 J. Loeb, *Proteins and the Theory of Colloidal Behavior*, New York, 1922.
17 R. O. Herzog and M. Kobel, *Z. Physiol. Chem.*, 134 (1924) 296.
18 N. Troensegaard and J. Schmidt, *Z. Physiol. Chem.*, 133 (1924) 116.
19 E. J. Cohn and J. B. Conant, *Z. Physiol. Chem.*, 159 (1926) 93.
20 G. S. Adair, *Proc. Roy. Soc. (London), Ser. A*, 109 (1925) 292.
21 T. Svedberg, *Kolloid Z.*, 51 (1930) 10.
22 A. Tiselius, *Nova Acta Regiae Soc. Sci. Upsaliensis*, 7 (1930) No. 4.
23 A. Tiselius, *Biochem. J.*, 31 (1937) 313, 1464.
24 J. R. Colvin, D. B. Smith and W. H. Cook, *Chem. Rev.*, 54 (1954) 687.
25 F. Sanger, in D. E. Green (Ed.), *Currents in Biochemical Research 1956*, New York, 1956.
26 M. O. Dayhoff, *Atlas of Protein Sequence and Structure 1969*, Silver Spring, Md., 1969.

Chapter 16

Biochemists Find Their Way into the Cell

1. The theory of protoplasm

Since Schwann formulated the concept of the "elementary parts" (the differentiated cells) as units of life, a view which has since acquired the status of a paradigm, a number of researchers have endeavoured to isolate something "living" beyond the cell.

An example was the theory of Huxley[1], stating that protoplasm was "the physical basis of life". The term protoplasm, designating the part of the cell outside the nucleus, was coined by Schultze. The nucleus, he says,

> "is surrounded by the cell substance proper, a viscous "protoplasm" which is non-transparent on account of it being densely filled with granules of a proteinaceous and fatty nature and which can be subdivided into a vitreous, transparent ground substance of a viscosity peculiar to protoplasm in general, and numerous imbedded granules." (Schultze[2], translation of Hall[3])

Schwann, when reading this definition, wrote in his lecture notes: "It is only another name for my "*Zellenschichte*"[4] (his definition of the cell was: a layer around a nucleus).

The philosophical implications of Huxley's attitude and that of his opponents has been studied by Hall[3] and by Geison[5]. It is an old tendency of prescientific biology to identify life with a special single substance as the prime locus of vital activity. Living particles were postulated by Democritus as well as by Buffon and, in the same category, we find the "primogenial moisture" of Fernel, the *materia vitae* of John Hunter, and, before the advent of "protoplasm", the "sarcode" of Dujardin. These philosophical theories have been adequately analyzed by Hall[3].

After the recognition of the cells as units of metabolism by Schwann, the old tendency towards identifying life with a single substance was revived in a number of theories. Many of them were concerned with hypothetical "metastructural" particles of "protoplasm", *i.e.* multimolecular but not

References p. 317

visible under the microscope: to this category belong the "physiological units" of Herbert Spencer, the "gemmules" of Darwin, the "plastidules" of Haeckel, the "plasson" of Ed. van Beneden, the "bioplasson" of Elsberg, the "micellae" of Nägeli, the "bioplasts" of Altmann, the "biophores" of Weismann, the "pangens" of De Vries, the "plasomes" of Hertwig, the "biogens" of Max Verworn. The history of all these ill-fated theories has been retraced by Hall[3], to whose book the reader is referred.

When the development of cytological techniques began to reveal a mass of new data on the structural aspects of cells, including many artifacts, several new theories of "protoplasmic" structure arose, among them some that pointed to a fibrillar or to a foam-like structure. The living-fibre theories of Ed. van Beneden and of Heitzmann and the foam theory of Bütschli are among those equally ill-fated theories of "protoplasmic" structure.

The view according to which the "protoplasm" is composed of a colloidal gel, *i.e.* presenting a gelatinous consistency and presumed to reveal the existence of colloids associated with water, was one of the results of the optical illusions afforded by the light microscope, and it reigned during the long barren period of biocolloidology referred to in Chapter 14.

The theory of Schwann, according to which the whole cell is the unit of life and, in biochemical terms, the unit of metabolism, has been supported by a number of leaders in biology and biochemistry and has become one of the leading concepts of modern biology.

In a book that has exerted a great impact on biological theory, Wilson[6] replied to the "metastructuralists" that "life can only be properly regarded as the property of the cell-system as a whole; and the separate elements could... better be designated as "active" or "passive" than as "living" or "lifeless"". Later, Hopkins stated the concept of the life of the cell as

> "the expression of a particular dynamic equilibrium which obtains in a polyphasic system. Certain phases of it may be separated... but life, as we instinctively define it, is a property of the cell as a whole, because it depends upon the equilibrium displayed by the totality of the coexisting phases." (ref. 7)

2. Cell metabolism and cell structure

The scope of biochemistry is to understand, at the molecular level, the system at work in the cell. It is clear that the extensive work of the enzymologists, isolating and crystallizing enzymes, made them accessible to studies on their chemical structure and mechanism of activity, and was of great interest for chemists interested in molecules and macromolecules. However, that work fell short of fully satisfying the biochemist whose quest is for the form of the integration of molecular structures and molecular activities unfolding themselves in the avenues of the cells.

Nevertheless, these chemical studies were critical in the unravelling of the pathways of metabolism (dealt with later in this History) including those of the biosynthesis of macromolecules, a knowledge which confirmed that, in their genesis, the cells are the result of a recognition, interaction and coordination of those biosynthesized macromolecules during cellular division and genetic recombination. The phrase of Hopkins relating the life of the cell "to the equilibrium displayed by the totality of the existing phases" of the cell adds the notion of equilibrium to the concept of a functional differentiation between the cell components, a concept which is as old as the cell theory itself and which was clearly enunciated by Schwann[8].

It has long been supposed that the integration of the cell depends on a definite spatial localization of the enzymes and that a cell, as well as an organism, could be integrated through the existence of several intracellular organelles, each of them differentiated, and all of them integrated. In other words, a relation is easily postulated between cell metabolism and cell structure. Among many examples we may cite the observation made by Krebs and Henseleit[9] who, having described the pathway of the "ornithine cycle" in liver cells, observed that liver extract was inactive and concluded that the metabolic pathway must be linked to structure. Such concepts of "structuration" were often expressed in vague terms (anisotropy, heterogeneity, polarization, etc.).

3. The end of the light microscope era

During the first decades of our century, it was realized with great concern that microscopical technique could not unravel the intracellular topography of the biochemical events taking place in cells. This was a direct consequence of the limitations of the power resolution of the light microscope. By the end of the 19th century, the light microscope had exhausted its technical possibilities, and for the first three decades of the 20th century, frustration remained present among microscopists. In spite of their ingenuity in devising methods of manipulating cells by microsurgery and multiplying histochemical methods, cytologists faced without success such problems as the

References p. 317

chemical structure of the cell organelles revealed by the microscope, the distribution of biochemical events in the cell and particularly the blank territory offered to the light microscope by the microscopically undifferentiated cytoplasm, a specially distressing aspect for lucid cytologists.

It must be emphasized here that the situation was not immediately alleviated by the introduction of the electron microscope.

The development of enzymological studies, while it permitted the identification of the enzymes involved and the reconstruction *in vitro* of the biochemical systems of metabolic pathways, did not lead to any solution of the specific problem of biochemistry. Destructive as they were, the enzymological methods were not able to provide information, or quantitative data, on either the location of the biochemical events in cells or the chemical architecture of the cell.

It is during this period of impotence that biocolloidology, as we said in Chapter 14, for lack of a better discipline, afforded to the biologist a pseudoscientific theory in which the search for deeper information on the relations of structure and function was alleviated in irrelevant theories related to surface actions, electric charges and adsorption.

But, since 1920, the recognition of proteins as chemically and physicochemically defined compounds, prepared the way for the fall of biocolloidology, as was shown in Chapter 15.

The cytologists, as said above, attempted to escape from this situation by trying to determine by specific colour tests, under the light microscope, the distribution of enzymes within the cell. Histochemical procedures were found to damage or destroy cell structures and to involve reactions incompatible with the life of cells. This derogatory statement does not apply to some of the reactions involved in histochemical methods. The benzidine and the indophenol reactions[10], for instance, do not merit these reproaches. The colour produced by these reagents at the level of oxidative systems in the presence of oxygen or H_2O_2 is found in discrete granules of the cytoplasm[11-13], but the products of the reaction (benzidine blue, indophenol blue) have a special affinity for the cytoplasmic granules, and they may be deposited in these granules secondarily after they have arisen elsewhere[14].

The lack of specificity of the micro-tests, as mentioned above, does not apply to the nuclear reaction of Feulgen which allows, under proper conditions, detection of the presence of DNA.

4. Differential centrifugation introduced by Claude as an analytical, quantitative method

The rupture that opened the modern field to the studies on the biochemical structuration of cells was accomplished by Claude who, in 1946, introduced the quantitative method of differential centrifugation to which he was led by a series of cancer studies*.

Claude had been interested in finding a method of separating organelles from cells. As a pre-medical student he had become interested in the eosinophilia as observed in cancers[15], and first tried to collect the eosinophile granules that are present in the sputum of asthmatic patients, but was unable to collect enough of them for study, partly owing to the high viscosity of the starting material.

In passing from pre-medical to clinical medical school (1924–1925), Claude worked out for himself a comprehensive programme, as a preliminary to the study of tumour cells. By that time the cell theory had been consolidated by two major observations. (*a*) Cell lines cultured *in vitro* were autonomous, able to grow and divide practically indefinitely, away and beyond the life spans of the organisms of origin. (*b*) Since the first successful grafting of tumour cells by Morau[16] and Jensen[17], and the subsequent painstaking work of the research group at the British Imperial Cancer Research Fund[18], it was demonstrated that tumour cells were not only autonomous; in addition they were independent of the growth control of their hosts. Claude realized that, as he put it later[33], "all the essentials of the living processes were to be found within the narrow boundaries of the cell wall". In order to familiarize himself with the problem, he decided to repeat the classical work of the British school, but with a variant: he used heterologous, instead of homologous, grafting. The mouse sarcoma S37 was obtained directly from J. A. Murray, the director of the Imperial Cancer Research Laboratories in London, and the reactions of the host tissues were investigated, after transplantation in rats[19,20]. The presentation of this work resulted in the award of a travel fellowship, spent in Berlin in 1928–1929. By then, the study of cancer was entering, however slowly, the experimental stage, and three main concepts were available to account for the malignant transformation of cells. (*1*) Microorganisms as carcinogens had been investigated since Pasteur's time with repeated negative results.

* Personal communication of Dr. Claude to the author.

References p. 317

Plate 58. Albert Claude.

However, the hypothesis was revived with the work of Erwin F. Smith on plant crown gall tumours[21]. (*2*) An encouraging field had been opened in 1911 by Rous[22] and Fujinami[23], by the demonstration of a cell-free, transmissible agent from chicken sarcoma, but the nature and rôle of this filterable agent was still highly controversial in the 1920's. (*3*) Growing evidence from different sources indicated that the malignant changes could be produced by a variety of physical agents, but this diversity and the remoteness of action were not appealing from the point of view of cell studies. Taking advantage of his fellowship, Claude decided to go to Berlin to acquaint himself with the work of Blumenthal, then Director of the Institut für Krebsforschung at Berlin University, who was claiming support for the bacterial theory from his experiments on the production of mammary tumours in mice, the technique being the inoculation of whole blood from mammary-tumour-bearing donors. The demonstration that the new growths resulted from the presence of viable tumour cells in the circulating blood of the bearer, along with non-pathogenic bacteria, brought to an early end this phase of cancer research. Claude spent the rest of his fellowship more to his liking, learning tissue culture with Albert Fischer, then at the Kaiser Wilhelm Institute in Berlin-Dahlem, and growing cells of the line of Sarcoma-37 that he had brought along with him. This experience with tissue culture was going to be useful years later, when new ways had to be found for the examination of cells by electron microscopy[26,42,★]. In the hope of continuing his programme, Claude wrote to Simon Flexner, Director of the Rockefeller Institute in New York, submitting a project aiming at the isolation of the filterable chicken tumour agent, and its further, chemical identification. He had expressed the wish to work in the laboratories of either Peyton Rous or Alexis Carrel, the two Departments where, to his knowledge, this tumour was being investigated. It turned out that Dr. Rous, who was more of a morphologist, had abandoned the cancer field for a number of years, and Carrel declined for lack of space. However, Dr. Flexner informed him that he would be welcome in another laboratory, headed by J. B. Murphy,

★ Specimens of Sarcoma-37 explants grown *in vitro* in Fischer's Laboratory and stained with giemsa, were on hand when, in 1933, Claude had the occasion to collaborate with the Mexican fresco painter, Diego Rivera, in providing him with the microscopic biological material which became part of the fresco "Man at the Crossroad" in the main lobby of the New York Rockefeller Center RCA Building: the S-37 tissue culture, with tumour cells streaming out, was used to balance the flaming sun opposite, in the telescopic field of the picture; the same fresco was reproduced later by the artist in the Hall of the Palacio de Bellas Artes, Mexico City, D. F. (Mexico).

References p. 317

where he soon found the possibilities to develop his own programme. He achieved the concentration and isolation of the agent, by fractionation, by adsorption and elution with alumina gel and by differential centrifugation[24]. The agent he showed to be a ribonucleic acid[25]. Later, when the electron microscope was available, he observed the particles of this agent in the tumour cells and proved it was a virus[26].

The recourse to differential centrifugation in the course of this highly significant work induced Claude to engage in a study of normal cells based on differential centrifugation. He had the idea of preparing a suspension of disintegrated cells in the way that cooks prepare a mayonnaise, and to submit it to differential centrifugation.

Claude himself recalls as follows the previous attempts to obtain components of cytoplasm.

"Warburg[27] was apparently the first to attempt to segregate cell components by means of centrifugation and to relate the biochemical activity of tissue extracts to preformed elements of the cytoplasm. Although the centrifugal force used by him was not sufficient to effect a complete separation of the large granules which, under the microscope, form a conspicuous part of mammalian liver extracts, a considerable concentration was obtained, and Warburg was able to show that the activity of these large elements was responsible for most of the oxygen uptake exhibited by cell-free extracts of guinea pig liver. Furthermore he convinced himself that the large granules of the extract were identical with those found in the cytoplasm of intact liver cells and pointed out that these elements could be seen leaving the cells during the preparation of a liver suspension. By means of a greater centrifugal force, Bensley and Hoerr[28] succeeded in separating the large granules of guinea pig liver more completely, and concentrates of large granules thus obtained were found by Lazarow[29] and Barron[30] to take up oxygen, and to possess succinoxidase activity." (ref. 31)

The mitochondrion affords a good example of an organized biochemical system (see this Treatise, Vol. 23). Mitochondria are cytoplasmic organelles that may be stained by crystal violet and that take janus green in vital coloration.

Benda, who introduced the crystal violet staining, coined the name *mitochondrion* (from *mitos*, a thread, and *chondros*, a grain), but Cowdry[32] has listed a number of other names applied to the same granules at one time or another (blepharoblasts, chondriokonts, chondriomites, chondrioplasts, chondriosomes, chondriospheres, fila, fuchsinophilic granules, interstitial bodies, Körner, Fadenkörner, Mitogel, parabasal bodies, plasmosomes, plastochondria, plastosomes, spheroplasts, vermicules, etc.). In the beginning of this century, a school of cytologists, led by Meves, proposed a theory in which mitochondria play a role in cell differentiation and heredity.

The modern era of research on respiration began with the definition of dehydrogenases, as we shall see in a later chapter. As pointed out by Lehninger,

> "actually very few biochemists concerned themselves with the possible importance of the fact that respiratory enzymes were found to be associated with particulate matter of cells and tissues. It was a part of the biochemical *Zeitgeist* that particles were a nuisance and stood in the way of purification of the respiratory enzymes." (ref. 33)

It was in 1940 that Claude[31], at the Rockefeller Institute in New York, started his pioneer work on the structure and composition of mitochondria (large granules) and microsomes (small granules) and founded the cytological school of the Rockefeller Institute.

The large granules isolated in Claude's pioneer work led him to recognize mitochondria as the "power plant" of the cell, by showing that cytochrome-linked enzymes and cytochrome *c* were localized in the "mitochondrial fraction". This work was extended at the Rockefeller Institute and this led, in 1948, to the isolation of intact mitochondria, by differential centrifugation in a medium of 0.88 *M* sucrose[34,35].

Claude, for the first time, detected submicroscopical particles, isolated by high-speed centrifugation[36,37], which he called "small particles" or "microsomes"[38,39].

It must be stressed here that these microsomes did not correspond to any optical images, either by light microscopy or electron microscopy. The discovery of submicroscopical units was based on differential centrifugation, and we owe to Claude the rupture with the era of microscopical study of cells, which had been paramount.

In his classical papers of 1946[31], which integrated more than ten years of systematic work on the subject, Claude offered for the first time a method of cell fractionation by differential centrifugation which separated cellular sub-units while preserving their structure and biochemical activities and made possible the studies *in vitro* of their morphology, composition and specific activities. One of the main aspects of Claude's contribution is that he insisted, from the start of his studies, that the method of mechanical fractionation

> "would afford a systematic inventory of the cell content, and a mapping of its biochemical activities, under the given condition, however, that it should be carried out on a strict quantitative basis and in such a way as to provide a complete balance-sheet of the results, which would express quantitatively the distribution of substances and of biological activities among the various cell fractions, and in per cents of those found in the whole cell, or in the unfractionated cytoplasm." (ref. 40)

References p. 317

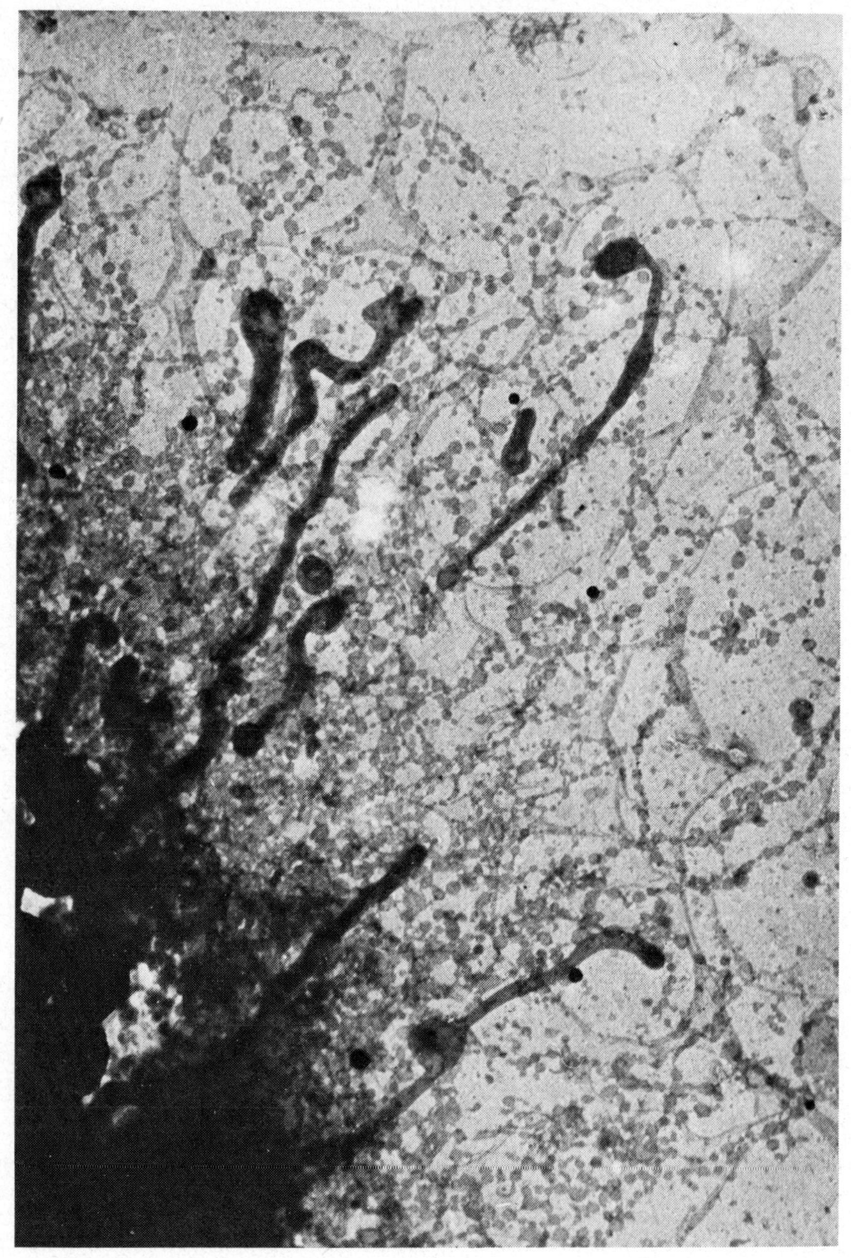

Plate 59. Transformation *in situ* of "endoplasmic reticulum" into "microsomes". (Claude[26,45])
Magnification: $\times$ 7500.

It is important to emphasize that, in their pioneer research, Miescher, for instance, endeavoured to separate pure nuclei and Bensley and Hoerr mitochondria (which they considered as "coacervates", within the frame of the current biocolloidology of the time) without aiming at any quantitative recovery, which obtained for the first time in Claude's methodology and gave to it the dimension of an analytical fractionation.

5. Microsomes and the endoplasmic reticulum

The "microsomes" isolated by Claude appeared, from his analysis, as constant components of the cells. Some considered them as dirt. Another suggestion was made, according to which they were immunoglobulins. Claude[41,42], in 1939, took pieces of *Amphiuma* liver, submitted them to centrifugation at $20\,000 \times g$ for one to two hours, then fixed, sectioned and stained them. The light microscope revealed a segregation of the various components within the cell, as they had been segregated in the tube of the centrifuge and in the same order. At the centrifugal pole, the glycogen was accumulated, and above it the nucleus and mitochondria appeared. Above these a wide zone was found, strongly basophilic as were the "microsomes", and structureless in the light microscope.

The relation of the "microsomes" with extant structures within the cell was elucidated later. Porter, Claude and Fullam[43], in the course of their early observations of cells with the electron microscope, described in 1945, in cytoplasmic areas, besides mitochondria and lipid droplets already known by light microscopy, a lace-like reticulum extending from the centre to the rim of the cell. Vesicle-like bodies, 100 to 150 mμ in size, were also observed along the strands of the reticulum, or scattered in the cytoplasm.

Claude, Porter and Pickels[26], in 1947, with recourse to an improved technique and to a better electron microscope (RCA, model C) assuring a superior resolving power, obtained a much better picture, as seen in Plate 59, of what they called the "cytoplasmic network", which was later given the name of "endoplasmic reticulum" by Porter[44] to indicate that in culture cells, the network is more concentrated in the endoplasm. Plate 59 is also interesting as illustrating the mechanism by which endoplasmic membranes may be transformed into vesicles and provide the "microsomes", the form of the reticulum utilizable *in vitro* for biochemical studies. In the left and the lower left corner of Plate 59 are seen vesicles of large size, next

Plate 60. George E. Palade.

to or in continuity with wider masses. This shows that in the process of fragmentation of endoplasmic membranes by vesiculation, there is a bulging, closing-up and sealing of portions and no leakage of content by breaking or opening up of the cavity bounded by the membrane. The "microsome" is therefore representative, biochemically, of the structure[45], a conclusion that accounts for the value of the biochemical data obtained with the artifact named "microsome". For electron microscopy therefore the microsomes represent segments of the endoplasmic reticulum transformed into vesicles. Palade[46] described their structure precisely by showing that nucleic acid is mainly concentrated in small grains of a diameter between 150 and 200 Å associated with the membranes. These grains, also found free in cytoplasm and responsible for its basophilia, were called *ribosomes* by Roberts[47].

The endoplasmic reticulum (ER), in plant cells as well as animal cells, including liver cells, is a system of interconnected channels lined by membranes whose framework is essentially composed of phospholipid bi-layers lined, at their hydrophilic faces, by layers of proteins. Two parts can be distinguished in this system: the rough ER, the surface of which is studded or lined with ribosomes and in which proteins and other substances are synthesized, including lipids needed for the synthesis of new membranes. The products may be transferred through the channels of the reticulum to other parts of the cytoplasm, to the nucleus, or to storage structures [lipoprotein vesicles, peroxisomes, products which are to leave the cell, are carried to its borders by the tubules or vesicles of the smooth ER, or they are carried by the same tubules to vesicules for storage (salivary gland, proteinase of exocrine pancreas) for later excretion].

The fine submicroscopical structure and the biochemistry of the endoplasmic reticulum was much advanced by the pioneer work of Palade and his associates, especially Siekevitz[48–52]. These studies were based on an integrated morphological (electron microscopy) and biochemical analysis of microsomal fractions and on cells (hepatocytes, exocrine cells of the pancreas) from which these fractions were derived. The results showed that the microsomes are membrane-bounded vesicles, most of them derived (by a generalized pinching-off process) from the ER. Those derived from the rough ER are marked by ribosomes still attached to their membranes. Free ribosomes were obtained by further centrifugation as "post microsomal fractions", and attached ribosomes were prepared by treatment of the microsomes with detergent.

References p. 317

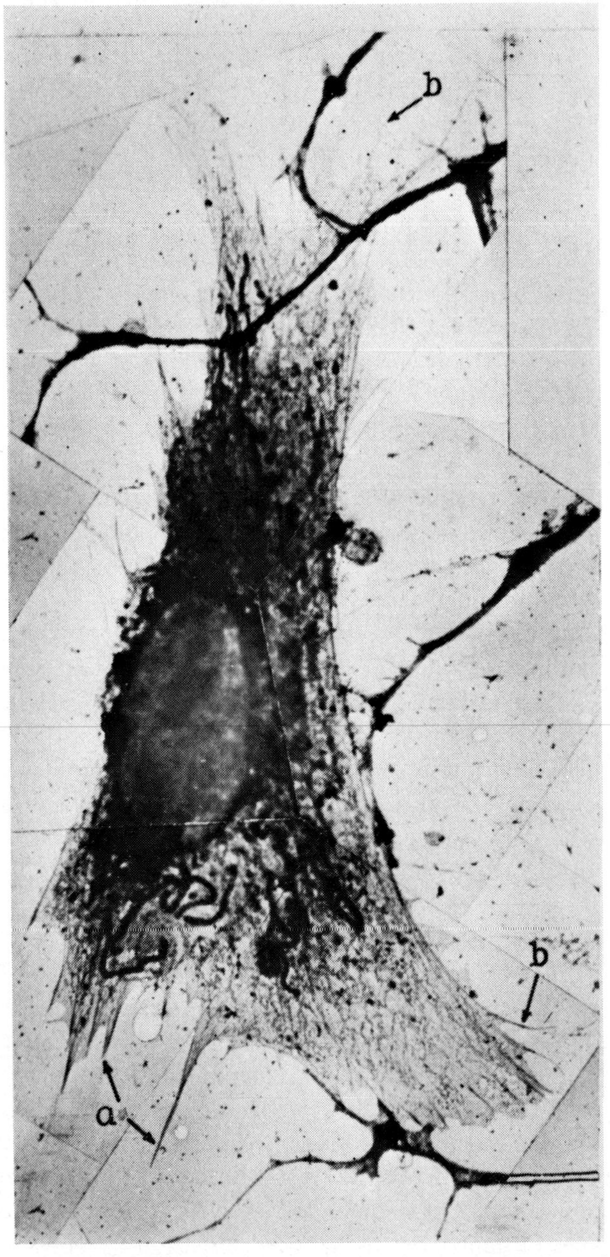

Plate 61. First composite picture of a chick embryo fibroblast-like cell in culture[26]. Photograph taken 6 July 1944. Mitochondria are clearly outlined, and a delicate reticulum is revealed. Magnification: ×1300. (Claude[42])

6. The contribution of electron microscopy to studies of cellular structure

Although an electron microscope had been conceived and built by Ruska[54] between 1930 and 1934, it is only since 1950 that the difficulties first encountered in its application to cell research (lack of contrast, excessive thickness of slices, defects in fixation, etc.) were progressively resolved. The limitations of the resolving power of the electron microscope first imposed limitations on its use. However, the thickness of the specimens posed a major problem. Organelles isolated by differential centrifugation were first examined. Another method for the preparation of specimens takes advantage of the flattening of cells in tissue cultures. The composite picture of Plate 61, a photograph taken by Claude[26] in New York on 6 July 1944 shows a chick-embryo fibroblast-like cell in culture. Mitochondria are clearly outlined, as is the cytoplasmic reticulum whose continuity is obvious.

In 1946, Claude and Fullam[54] constructed a first ultramicrotome and accomplished electron microscopy on tissue sections. In this instrument, the cutting blade revolved at high speed. The problem of handling ultrathin sections had not been resolved at the time, and the sections blown from the blade were received on a wire mesh. In 1947, Claude designed a new microtome, experimental in purpose, in which the main parts could be oriented and maintained at various angles, and were interchangeable. It is on this occasion that Claude introduced for the first time two devices essential in the making of thin sections: (*1*) the bypass which avoided the upward stroke of the blade or of the block, and (*2*) the so-called water trough which made it possible to receive the section directly on to a liquid surface where it could expand to its original size, thanks to the surface tension of the fluid[42]. This instrument (Plate 62) was used by Palade in his classical work on fixation for electron microscopy.

The productive method of the investigation by electron microscopy dates from a paper of Palade[56] published in 1952. The fixation techniques described in this paper (OsO_4 in a medium buffered at pH 7.4) were of paramount importance in the developments of the following ten years.

One of the first important accomplishments of electron microscopy, and which is due to Palade, concerns the structure of the endoplasmic reticulum and has been referred to above. Another outcome of the technical improvements introduced by Palade is represented by his pioneer studies on the microscopical structure of mitochondria[57,58] in which the "cristae

References p. 317

Plate 62. Claude–Blum microtome[42].

mitochondriales" were first described as was the relationship of the two mitochondrial membranes with one another, and the inner membrane with the cristae was defined. The location of the respiratory chains in the inner membranes and the cristae was also postulated[58]. (The mitochondrial structure was investigated independently at about the same time by F. S. Sjöstrand.)

This work was at the origin of the rapid development of our knowledge of mitochondrial structure (see Vol. 14 of this Treatise) which led to the correlation of mitochondria with the metabolic systems of the Krebs cycle and the respiratory chains, and with fatty acid oxidation (see Part III of this History).

7. Lysosomes

In the year 1949, de Duve, in his laboratory at Louvain in Belgium, was interested in a hexose phosphatase first discovered in the liver by Cori and Cori. De Duve's intent was oriented to the mechanism of insulin's action, and he suspected that the hexose phosphatase could act as a direct antagonist of the hexokinase and mask the action of insulin *in vitro*. In the course of the purification of the enzyme from rat-liver homogenates by his colleague Hers, it appeared that it could be precipitated at pH 5 but could not be redissolved at any pH between 5 and 9.

"By some fortunate coincidence" [mentions de Duve in telling this story] "my recent readings had included the two now classical, papers by Albert Claude[31] on "Fractionation of mammalian liver cells by differential centrifugation" as well as the subsequent description by Hogeboom *et al.*[35] (1948)[26] of "The isolation of morphologically intact mitochondria from rat liver". These papers had made a deep impression on me because they opened the possibility of exploring cellular organization by biochemical methods. When Hers reported his failure to dissolve the precipitated glucose 6-phosphatase, I immediately recalled Claude's diagrams showing the agglutination at pH 5 of both large and small granules and concluded that our enzyme was likely to be firmly attached to some kind of subcellular structure." (ref. 59)

A high speed head was therefore acquired for the laboratory's centrifuge and, de Duve writes,

"We were also greatly helped by Claude, who had just returned to Belgium and who taught us a number of valuable tricks, such as the "mayonnaise" method for resuspending sediments. No doubt his explanation would have been lost on any but continental amateur cooks." (ref. 59)

The enzyme G-6-phosphatase was found almost entirely sedimentable, more than two-thirds of the activity being recovered in the microsome

References p. 317

Plate 63. Christian de Duve.

fraction and the remainder distributed between the nuclear and mitochondrial fractions.

In experiments performed with a better technique, Hers *et al.*[60] succeeded in isolating 95% of the enzyme with the microsome fraction. This demonstration of the unique localization of G-6-phosphatase in microsomes, together with the work of Hogeboom *et al.*[61] showing the unique localization of cytochrome oxidase in mitochondria, was of importance in the definition of biochemical homogeneity and of unique localization both of which were of paramount importance in later researches.

The ultracentrifugation of the unspecific acid phosphatase was far from being so clear. In the course of the relevant studies with fresh preparations, the enzyme was found in the mitochondrial fraction. But when aged mitochondrial fractions were centrifuged, most of the acid phosphatase activity remained in the supernatant. This example of structure-linked latency became the main study of de Duve and his colleagues, and a few months later it was recognized that the latency was the expression of the presence of the enzyme in a sac-like structure surrounded by a membrane (see de Duve[59]). This structure was later called "lysosome"[55], after it was found to contain several other acid hydrolases. It must be stressed that the approach of de Duve was of an exclusively biochemical and analytical nature. Lysosomes were first identified as families of sedimentable enzymes, the acid hydrolases. In fractionation experiments, repeated with different methods, very similar distributions were revealed, and at the same time they appeared to be significantly different from the distributions of enzymes known as mitochondrial or microsomal. This biochemical method can only be a fruitful way of enquiry if each of the organelles of a cell has a definite enzymatic composition: this is what de Duve has called the *postulate of biochemical homogeneity*. Another postulate he assumes, though being ready for exceptions, is that of *single location*, that is, each enzyme is restricted to a definite site in the cellular structure. These two postulates, clearly formulated by de Duve[62–64], are of great theoretical value.

It is remarkable that, as de Duve explained in his Harvey Lecture[62], the behaviour of the enzymes in different conditions of centrifugation has led to determination of the density and dimensions of the granules, while the consideration of the enzymatic latency has led to the recognition that the granules were surrounded by a membrane and to knowledge of the physical and chemical characteristics of that membrane. The existence of the lysosomes was therefore recognized by de Duve through a biochemical

References p. 317

approach, comparable to the approach of Leverrier in concluding from his calculations the existence of the planet Neptune.

As Neptune was looked for by Galle who found it through his telescope, the lysosomes were recognized in electron-micrographs of a preparation Novikoff obtained from De Duve when he visited Brussels at the time of the 3rd Biochemical Congress[65].

8. Peroxisomes

Another demonstration of the fecundity of the method of analytical differential centrifugation is afforded by the discovery of peroxisomes. In 1953, Novikoff *et al.*[66] found that, in rat liver, urate oxidase shows sedimentation properties similar to those of acid phosphatase. As at that time de Duve and his colleagues were busy with the study of lysosomes they confirmed the observation of Novikoff, but they observed that urate oxidase has neither the solubility nor the latency properties of the lysosomal hydrolases. Two other non-hydrolytic enzymes, D-amino acid oxidase[67] and catalase[68], were found to sediment at the same rate as the acid hydrolases and urate oxidase, but these two enzymes, in density-gradient experiments, follow urate oxidase rather than the acid hydrolases. (Literature in de Duve and Baudhuin[69] where the reader will find an elaborate account of the work on peroxisomes.)

Since that time it has been established by combined biochemical and morphological studies that the particles containing the enzymes concerned correspond to what was described in 1954 by Rhodin[70], who used the electron microscope, as special granules in the convoluted tubule of the mouse kidney and called "microbodies". The name *peroxisome* was coined by de Duve[71], in his biochemical perspectives, to indicate that these particles are the site of hydrogen peroxide metabolism in which an association of oxidases and catalases has a biochemical significance.

Contrary to mitochondria, most of which are situated outside the smooth region and are free of contact with smooth membranes, peroxisomes preferentially occur in regions of the smooth ER lamellae and the Golgi region[69], which suggests the possibility of exchanges between peroxisomes and smooth ER. De Duve suggests the existence, in the peroxisomes, of an auxiliary metabolic mechanism which could be of use in the disposal of hydrogen peroxide and in several metabolic pathways in which molecules susceptible to peroxidation by catalase are concerned. This mechanism could also be of importance in the system of gluconeogenesis (see de Duve and Baudhuin[69]). Recent results indicate that peroxisomes play a particularly important role in plant cells.

9. Principles of tissue fractionation

The analytical method of differential centrifugation is based on a number of premises that have been developed logically by de Duve. He stresses that the best approach is a purely analytical one

> "untrammeled by any preconceived idea of the cytological composition of the isolated fractions, and allowing their biochemical properties to be expressed as continuous functions of the physical parameter which determines the behaviour of subcellular components in the fractionation system chosen." (ref. 64)

In this type of approach, great advantage may be obtained from the density-gradient centrifugation, as the degree of separability of several groups of particles varies with the composition of the medium. In this analytical approach, the independent variable is related "to a property, not of the constituent, but of its host-particles".

> "In the design of tissue fractionation experiments and in the interpretation of their results, it is essential to observe a rigorous logic and to employ an appropriately accurate vocabulary. The most important requirement in this respect is to maintain a strict distinction between the intracellular organelles or structures as they occur within the cells, the population of particulate aggregates as they are present in the homogenate and react to the fractionation procedure, and the subcellular fractions as they are isolated and analyzed." (de Duve[64])

10. Synthesis of lipoprotein granules

It has been recognized for some time, by biochemists and physiologists, that so-called low (LDL) and very-low density lipoproteins (VLDL) of blood plasma have their origin in liver. It is only in recent years, starting in about 1964 (Novikoff[72]), that this field has benefited from observations of liver by electron microscopy. The most significant contributions in this respect have been from the works of Baglio and Farber[73], Lombardi *et al.*[74–76], Jones *et al.*[77], Hamilton *et al.*[78], Stein and Stein[79], and Claude[80]. The observations of these authors were made on liver, *in vivo* or perfused, by quantitative measurements with radioactive lipid precursors, including autoradiography, on the sites of synthesis, turnover and transport of plasma lipoproteins of hepatic origin. The actual lipoprotein nature of these lipids, which have high triglyceride content and present themselves in the form of discrete granules of 350 Å or more in diameter in liver and in plasma, is far from being substantiated. Likewise, it is not clear whether

References p. 317

all, or any, of the lipoprotein granules which reach the plasma do it by way of the Golgi apparatus, or are carried to the blood capillaries directly by the smooth endoplasmic reticulum.

The recent observations of Claude on the subject[80,81] have reopened the question of the disposal of the lipoprotein granules and of the significance of this metabolic process in terms of the general physiology of the hepatic cell. Claude described a new, apparently specific, pathway where an important part of the synthesized lipoprotein granules are concentrated, from the smooth endoplasmic reticulum into the Golgi apparatus, then digested through the fusion of coated vesicles with concentrating Golgi sacs, and recycled into the hepatic cell metabolism, with possible involvement in the process of bile formation (Claude[80,81]). The Golgi apparatus in liver is characteristically localized in the vicinity of the bile canaliculi.

11. Biochemical structure of cells

The approaches defined above by their conceptual and methodological components have led to the collection of a large number of data on the composition and structure of the different cell organelles and on their metabolic interrelations. (see Vol. 23 of *Comprehensive Biochemistry* for the development of knowledge concerning nucleus, nucleolus, cell membrane and lysosome; also Vol. 14 for the structure of the mitochondrion.)

At the level where biochemical and electron-microscopical studies meet in this development, when structure is synonymous with assembly of macromolecules and function is related to the activities of the same macromolecules, the proper scope of biochemistry eventually reaches its primary and promising accomplishments.

It must be stressed that in this perspective, the cell, the living unit, itself becomes an organism, the study of which is exposed to the same pitfalls as those the first physiologists met when the whole organism imposed itself on their sensorial perception in the form of *anatomia animata*. The frequent use, in recent times, with reference to cells, of such phrases as "biological machines", or of "molecular technology" indicates that the pipes, sieves and levers of the iatrophysicist and his naive biophysics are already around the corner, ready to take their place in a new "cell biology" of a descriptive anthropomorphic nature and to postpone the development of a more abstract scientific treatment of the cellular polyphasic system of integrated macromolecules, commonly known as life.

REFERENCES

1 T. H. Huxley, *Fortnightly Rev.*, 5 (1869) 129.
2 M. Schultze, *Arch. Anat. Physiol. Wiss. Med.*, (1861) 1.
3 T. S. Hall, *Ideas of Life and Matter*, Vol. 2, Chicago and London, 1969.
4 M. Florkin, *Naissance et Déviation de la Théorie Cellulaire dans l'Oeuvre de Théodore Schwann*, Paris, 1960.
5 G. L. Geison, *Isis*, 60 (1969) 273.
6 E. B. Wilson, *The Cell in Development and Heredity*, New York, 1925.
7 F. G. Hopkins, *Nature*, 92 (1913) 220.
8 Th. Schwann, *Mikroskopische Untersuchungen*, Berlin, 1839.
9 H. A. Krebs and K. Henseleit, *Z. Physiol. Chem.*, 210 (1932) 33.
10 L. Lison, *Histochimie Animale*, Paris, 1936.
11 E. Gierke, *Münch. Med. Wschr.*, 11 (1911) 2315.
12 S. Graeff, *Frankf. Z. Pathol.*, 11 (1912) 368.
13 S. Graeff, in E. Abderhalden (Ed.), *Handbuch der biologischen Arbeitsmethoden*, Abt. 4, Teil 1, Heft 1, Lief. 78, Berlin and Vienna, 1922.
14 A. C. Hollande, *Compt. Rend.*, 178 (1924) 1215.
15 A. Claude, *Liège Méd.*, No. 23 (1929) 1.
16 H. Morau, *Compt. Rend. Soc. Biol.*, 43 (1891) 289.
17 C. O. Jensen, *Centralbl. Bacteriol.*, Abt. I, 28 (1903) 122.
18 J. A. Murray, *Third Scientific Report Imperial Cancer Research Fund*, London, 1908, 175.
19 A. Claude, *Compt. Rend. Soc. Biol.*, 99 (1928) 1058.
20 A. Claude, *Compt. Rend. Soc. Biol.*, 99 (1928) 1061.
21 E. F. Smith, *J. Cancer Res.*, 1 (1916) 231.
22 P. Rous, *J. Am. Med. Ass.*, 56 (1911) 198.
23 A. Fujinami and K. Inamoto, *Z. Krebsforsch.*, 14 (1914) 94.
24 A. Claude, *Science*, 87 (1938) 467.
25 A. Claude, *Science*, 90 (1939) 213.
26 A. Claude, R. K. Porter and E. G. Pickels, *Cancer Res.*, 7 (1947) 421.
27 O. Warburg, *Arch. Ges. Physiol.*, 154 (1913) 599.
28 R. R. Bensley and N. L. Hoerr, *Anat. Record*, 60 (1934) 449.
29 A. Lazarow, *Biol. Symposia*, 10 (1943) 9.
30 E. S. G. Barron, *Biol. Symposia*, 10 (1943) 27.
31 A. Claude, *J. Exptl. Med.*, 84 (1946) 51, 61.
32 E. V. Cowdry, *Contrib. Embryol.*, 8 (1918) 39.
33 A. L. Lehninger, *The Mitochondrion, Molecular Basis of Structure and Function*, New York, 1964.
34 G. H. Hogeboom, A. Claude and R. D. Hotchkiss, *J. Biol. Chem.*, 165 (1946) 615.
35 G. H. Hogeboom, W. C. Schneider and G. H. Palade, *J. Biol. Chem.*, 172 (1948) 619.
36 A. Claude, *Proc. Soc. Exptl. Biol. Med.*, 39 (1938) 398.
37 A. Claude, *Science*, 91 (1940) 77.
38 A. Claude, *Science*, 97 (1943) 451.
39 A. Claude, *Biol. Symposia*, 10 (1943) 111.
40 A. Claude, in P. Buffa (Ed.), *From Molecule to Cell: Symposium on Electron Microscopy*, Rome, 1964.
41 A. Claude, in *Biol. Symposia*, 10 (1943) 9 (see also plate 9 in ref. 42).
42 A. Claude, *Harvey Lectures*, 43 (1947–1948) 121.
43 K. R. Porter, A. Claude and E. F. Fullam, *J. Exptl. Med.*, 81 (1945) 233.
44 K. R. Porter, *J. Exptl. Med.*, 97 (1953) 727.

45 A. Claude, in, *Microsomes and Drug Oxidation*, New York, 1969, p. 1.
46 G. E. Palade, *J. Biophys. Biochem. Cytol.*, 1 (1955) 59.
47 R. B. Roberts, in R. B. Roberts (Ed.), *Microsomal Particles and Protein Synthesis*, Oxford, 1958, p. VIII.
48 G. E. Palade and P. Siekevitz, *J. Biophys. Biochem. Cytol.*, 2 (1956) 171, 671.
49 L. Ernster, P. Siekevitz and G. E. Palade, *J. Cell Biol.*, 15 (1962) 541.
50 G. Dallner, P. Siekevitz and G. E. Palade, *J. Cell Biol.*, 30 (1966) 73, 97.
51 P. Siekevitz, G. E. Palade, G. Dallner, I. Ohad and T. Omura, in H. J. Vogel, J. O. Lampen and V. Bryson (Eds.), *Organizational Biosynthesis*, New York, 1967, p. 331.
52 J. D. Jamieson and G. E. Palade, *J. Cell Biol.*, 34 (1967) 577, 597.
53 E. Ruska, *Z. Phys.*, 87 (1934) 580.
54 A. Claude and E. F. Fullam, *J. Exptl. Med.*, 83 (1946) 499.
55 C. de Duve, B. C. Pressman, R. Gianotti, R. Wattiaux and F. Appelmans, *Biochem. J.*, 60 (1955) 604.
56 G. E. Palade, *J. Exptl. Med.*, 95 (1952) 285.
57 G. E. Palade, *Anat. Record*, 114 (1952) 427.
58 G. E. Palade, *J. Histochem. Cytochem.*, 1 (1953) 188.
59 C. de Duve, in J. T. Dingle and H. B. Fell (Eds.), *Lysosomes in Biology and Pathology, Vol. 1*, Amsterdam, 1969. p. 1.
60 H. G. Hers, J. Berthet, L. Berthet and C. de Duve, *Bull. Soc. Chim. Biol.*, 33 (1951) 21.
61 G. H. Hogeboom, W. C. Schneider and M. J. Striebich, *J. Biol. Chem.*, 196 (1952) 111.
62 C. de Duve, *Harvey Lectures*, 59 (1965) 49.
63 C. de Duve, in D. B. Roodyn (Ed.), *Enzyme Cytology*, New York, 1967, p. 1.
64 C. de Duve, *J. Theoret. Biol.*, 6 (1964) 33.
65 A. B. Novikoff, H. Beaufay and C. de Duve, *J. Biophys. Biochem. Cytol.*, 2 Suppl. (1956), 179.
66 A. B. Novikoff, E. Podber, J. Ryan and E. Noe, *J. Histol. Cytochem.*, 1 (1953) 27.
67 K. Paigen, *J. Biol. Chem.*, 206 (1954) 945.
68 J. F. Thomson and F. J. Klipfel, *Arch. Biochem. Biophys.*, 70 (1957) 224.
69 C. de Duve and F. Baudhuin, *Physiol. Rev.*, 46 (1966) 323.
70 J. Rhodin, *Aktiebolaget Godvil*, 1954 (cited after De Duve and Baudhuin[69]).
71 C. de Duve, *J. Cell Biol.*, 27 (1965) 25A.
72 A. B. Novikoff and W. Y. Shin, *J. Microscoscopy*, 3 (1964) 187.
73 C. M. Baglio and E. Farber, *J. Cell. Biol.*, 28 (1965) 277.
74 B. Lombardi and A. Oler, *J. Lab. Invest.*, 17 (1967) 308.
75 F. F. Schlunk and B. Lombardi, *J. Lab. Invest.*, 17 (1967) 30.
76 G. Ugazio and B. Lombardi, *J. Lab. Invest.*, 14 (1965) 711.
77 A. L. Jones, N. B. Rudeman and M. G. Herrera, *J. Lipid Res.*, 8 (1967) 429.
78 R. L. Hamilton, D. M. Regen, M. E. Gray and V. S. Lequire, *J. Lab. Invest.*, 16 (1967) 305.
79 O. Stein and Y. Stein, *J. Cell Biol.*, 33 (1967) 319.
80 A. Claude, *J. Cell Biol.*, 47 (1970) 715.
81 A. Claude, *Proc. VII Intern. Congr. of Electron Microscopy*, Grenoble, 1970, p. 85.

Appendices

Appendix 1

Origins of the International Union of Biochemistry

Since 1927, the Société de Chimie Biologique has organized regular congresses (Lille, 1927; Paris, 1929; Strasbourg, 1931; Paris, 1933; Brussels, 1935; Lyon, 1937; Liège, 1946; Paris, 1948).

Just after the second world war, when biochemists were again in a position to come together and resume a collaboration at the international level, the 7th Congress of the Société de Chimie Biologique, which took place in Liège (1946) was the occasion for the biochemists from several countries who attended it, to become conscious of the desirability of organizing international Congresses of Biochemistry.

An international Congress of Physiology was scheduled for the next year at Oxford, and the Biochemical Society, which has been in existence since 1911, approached the organizers "to ensure that Biochemistry was allocated its share in the programme"[1]. The reply stated that it was

> "impossible to issue a general invitation to biochemists to participate in the Congress and that, while no actual embargo would be placed on biochemical papers, these would have to come from, or be introduced by, members of the Physiological Society." (Morton[1])

Consequently the Biochemical Society took the first steps to organize an International Congress of Biochemistry. It must be stressed here that no hostility was expressed by the Physiological Society against the idea. The Committee of the Physiological Society, when it was made aware of the intention of the Biochemical Society to organize an International Congress of Biochemistry, maintained its position, adding that

> "if the Biochemical Society decided to initiate Congresses of their own they would have the Physiological Society's blessing, encouragement and offer of assistance." (Morton[1])

In the meantime an 8th Congress of the Société de Chimie Biologique was held in Paris in 1948. Invitations were sent to biochemists of different countries and discussions were planned on an international basis.

References p. 325

The Biochemical Society then made a preliminary enquiry about the desirability of organizing international congresses of biochemistry. Its Committee sent a circular to 100 biochemists (societies or individuals) asking their advice. Although it is true that only 24 answers were received, it must be emphasized that they were all favourable[2]. It is in this sense that we may interpret the phrase of Chibnall as president of the 1st Congress of Biochemistry, that the answer had been

"so favourable that the Society decided, at its Annual General Meeting in March 1948, to promote a Congress in Cambridge in August 1949." (Chibnall[2])

It may have been thought that the Cambridge Congress had no more international a nature than the previous French Congresses but the president, in his introductory address, mentioned that

"without commitment for the future, the project was officially recognized by the International Union of Chemistry, so that the First Congress of Biochemistry has official international backing. Following the usual custom, the Biochemical Society nominated a Congress Committee exclusively from the host country; the Congress Committee itself has taken the necessary steps to ensure that a small group of accredited representatives from various leading countries—under the chairmanship of Sir Charles Harington—shall use the excellent opportunity, afforded for full discussion at the Congress, to reach if possible some decision as to the means whereby the machinery necessary for organizing future Congresses can be established." (Chibnall[2])

While this Congress (19–25 August 1949) was called the "First International Congress of Biochemistry", Chibnall, in his opening address, remarked:

"For our French colleagues this will be the ninth Congress of Biochemistry; for the world as a whole, however, it will be the first to have been organized with full international machinery. We feel no doubt, therefore, that our friends from France will agree that this is properly designated the First International Congress of Biochemistry, not in forgetfulness of their past efforts, but rather with the full recognition that it is to those efforts and to the inspiration and encouragement which they have given that any success which we may achieve on the present occasion will be largely due." (Morton[1])

The three Congress general lectures were chosen to map the field of interests of biochemistry. One was on the organic chemistry of natural substances, but delivered by an organic chemist (Sir Robert Robinson on "Tryptophan and its structural relatives"). The two others were respectively on enzymology and metabolic regulation (C. F. Cori on the "Influence of hormones on enzymatic reactions") and on the biochemical aspects of general biology (M. Florkin on "Biochemical aspects of some biological concepts").

In the final session, Chibnall thanked the biochemists of all countries who had accepted, from "the biochemists of Great Britain" the invitation to come to this Congress. He told the audience that the small group of representatives of different countries, under the chairmanship of Sir Charles Harington, which he had mentioned in his opening talk, had met and had prepared some resolutions to put before the assembly. Sir Charles Harington then reported that his committee proposed:

(*1*) "That the next Congress be held in Paris in 1952, (*2*) "That an *International Committee for Biochemistry* be now set up; and that it have the following composition: *Chairman*, Sir Charles Harington; *Australasia*, A. H. Ennor; *Belgium*, M. Florkin; *Canada*, J. B. Collip; *France*, J. Roche and J. Courtois; *Great Britain*, H. Raistrick and J. N. Davidson; *India*, B. C. Guha; *Italy*, P. Pratesi; *The Netherlands*, H. G. K. Westenbrink; *Poland*, T. Baranowski; *Scandinavia*, K. Linderstrøm-Lang and A. Tiselius; *South America*, E. Cruz-Coke; *Switzerland*, W. H. Schöpfer; *United States*, R. W. Jackson and J. T. Edsall; *USSR*, two delegates recommended by the appropriate authority of the USSR. And that this Committee be empowered to co-opt additional members from other countries as may be desirable. (*3*) "That this Committee be instructed to approach the International Council of Scientific Unions with a request for recognition as the international body representative of biochemistry, with a view to the formal constitution of an International Union of Biochemistry as soon as possible." (Chibnall[2])

The resolutions were carried by acclamation.

But if the International Committee for Biochemistry set up in Cambridge in 1949 met with no difficulty on the part of the physiologists, this was far from being true for the chemists.

Early in 1949, before the First International Congress of Biochemistry met in Cambridge, the (British) National Committee for Chemistry had already resolved

"that the proposal for an International Union of Biochemistry would be better replaced by a proposal to establish a joint Committee between the International Union of Biological Sciences and the International Union of Chemistry which should be its mother union." (Morton[1])

In the same year the International Union of Pure and Applied Chemistry (IUPAC) decided, during its meeting at Amsterdam, that it would be composed of six sections, one of which would be devoted to Biological Chemistry under the chairmanship of A. W. K. Tiselius.

The International Committee for Biochemistry sent an application to ICSU concerning the establishment of an International Union of Biochemistry, and when the Council met in New York in September 1951, F. Dickens and J. N. Davidson went there to speak in its favour. At this meeting, Tiselius was elected President of IUPAC, and consequently the chairmanship of the Biological Chemistry section became vacant. Sir Charles Dodds accepted it "only on condition that it was understood and minuted that he was in favour of an independent Union of Biochemistry

References p. 325

and that he would continue to further the cause of an independent Union" (ref. 1). The International Committee for Biochemistry was asked to nominate five persons to fill vacancies in the Biological Chemistry section of IUPAC.

New vigour was given to the movement in favour of an independent Union of Biochemistry when Sir Rudolph Peters, in 1952, became chairman of the Committee of the Biochemical Society.

In July 1952, the International Committee on Biochemistry met in Paris, during the 2nd International Congress of Biochemistry. As Sir Charles Harington had resigned from the chairmanship, the chair was taken by Davidson, and the present author was, *in absentia*, elected chairman of the International Committee of Biochemistry. The new chairman, who had clearly realized, as a member of the Committee of the Section of Biological Chemistry of IUPAC, that it was hopeless to believe that the chemists would understand and promote the objectives of biochemistry, did not agree with the policy of going to ICSU for advice on the creation of a Union when nothing but opposition could be met from the representatives of IUPAC in the Council. But nevertheless, in September 1952 E. Brand (U.S.A.), H. G. K. Westenbrink (The Netherlands), J. N. Davidson (Great Britain) and M. Florkin (Belgium) went to Amsterdam to discuss the matter with the Council members, with no more result than before. It is clear that ICSU is a Council for federating International Scientific Unions, but it never was this writer's understanding that the creation of a Union should be subordinated to the blessing of ICSU. We therefore pushed the pioneer work ahead and by the beginning of 1953 applications from "adhering bodies" from 12 countries had been received. The International Committee of Biochemistry decided, on 1 March 1953, to found a permanent organism which would bear the name "International Union of Biochemistry". An interim Council of the Union was set up, chaired by the present author. On 5 and 6 January 1955, the First General Assembly of the International Union of Biochemistry met in the Senate House of the University of London under his chairmanship. By that time the "adhering bodies" of 15 countries had joined and 12 were represented. The statutes were formally adopted, and the first Council of IUB was elected as follows[3]: *Australia*, A. H. Ennor; *Austria*, O. Hoffmann-Ostenhof; *Belgium*, M. Florkin (President); *Denmark*, K. Linderstrøm-Lang; *Finland*, A. Virtanen; *France*, J. E. Courtois, J. Roche; *Germany*, K. Felix, K. Lohmann; *Great Britain*, Sir Rudolph Peters, Sir Charles Harington (Hon. Member),

R. H. S. Thompson (Secretary-General); *Italy*, A. Rossi-Fanelli; *Japan*, S. Akabori; *The Netherlands*, H. G. K. Westenbrink; *Norway*, R. Nicolaysen; *Sweden*, E. Hammarsten; *U.S.A.*, S. Ochoa; E. H. Stotz (Treasurer); *U.S.S.R.*, V. A. Engelhardt, A. I. Oparin.

When the Second General Assembly of IUB met in Brussels in August 1955, five further countries had been admitted (Canada, Czechoslovakia, Hungary, Poland, Spain)[3]. When the 7th General Assembly of ICSU met during the same month in Oslo, the President and Secretary-General of IUB went to defend its application for membership in ICSU, which was accepted.

One of the concerns of the first president of IUB was to cooperate with other Unions on matters of common interest. Coordinating committees were set up with the International Union of Pure and Applied Chemistry (which later, in 1967, dissolved its Division of Biological Chemistry) with the International Union of Physiological Sciences (IUPS) and later with the newly formed International Organization of Pure and Applied Biophysics (IOPAB). Cooperation with the International Union of Biological Sciences (IUBS) was ensured by the designation of the first President of IUB as first President of the Section of Biochemistry of IUBS.

The International Union of Biochemistry has supported the coopting in ICSU of the Union of Pure and Applied Biophysics and of the International Union of Nutrition, now members of ICSU. A joint IUB/IUPAC Biochemical Nomenclature Commission has been established and is active.

The successive Presidents of IUB have been: M. Florkin, K. Linderstrøm-Lang, S. Ochoa, H. Theorell.

REFERENCES

1 R. A. Morton, *The Biochemical Society, Its History and Archives 1911–1969*, London, 1969.

2 A. C. Chibnall, in *First International Congress of Biochemistry, held at Cambridge 19–25 August 1949. Report of Opening and Concluding Sessions and Three Lectures delivered at the Congress*, Cambridge, 1950.

3 R. H. S. Thompson, *ICSU Rev. of World Science*, 5 (1963) 142.

Appendix 2

On Alchemical Symbolism

Chapter 2, being devoted to the proto-biochemical aspects of alchemy and its rôle as precursor of iatrochemistry, does not pretend to give an adequate picture of all forms of alchemy or of its symbolism. We have traced the origins of Greek alchemy back to stoic philosophy, and it is believed that the originators of this variety of alchemy had connections with the gnostics, although this aspect is not documented enough and needs much more study. For the adepts of the gnostic heresy, knowledge was afforded by revelation only, which was not the creed of European alchemists. Their taste for symbolism and allegory may find its origins in gnostic attitudes, but we may recall that the same tendency was present in the stoic school.

Whatever theory we may adopt concerning the origin of its symbolism, when alchemy was imported from the Arabian world into Western Europe in the 12th century, this symbolism, in many of its aspects, entered into the Christian creed. As has been stressed by Sheppard,

"The esoteric side of European alchemy consists of a vast edifice of symbolism representing the need for a Christian redemption of man. Gone is the pagan background...; in its place is a Christian one, in which most of the gnostic symbolism has undergone a transformation to suit the prevailing religious moods." (ref. 1)

The "magnum opus", the operation in which the metals were treated to change their "form", led to a number of symbolic expressions. The distinction of the stages of the operation derives from colour changes supposed to be observed by hellenistic alchemists during the treatment of a material by heat. One example is the black state ("melanosis", "nigredo", producing the "prima materia") observed when copper is heated with sulphur and consequently a black copper sulphide is observed. This is followed by "leukosis", "albedo" or whitening when the material is further heated and consequently desulphurated. The following stages are "xanthosis" or yellowing, and "iosis" "rubedo" or reddening.

In the late Middle Ages, the alchemists were not referring to connections between different fields of knowledge, as we do, but to connections between the domains of the *psyche*, relating the stages of calcination to the Christian symbolism of passion, death and resurrection. "Nigredo" corresponds to the passion and mortification of Christ and it was followed by a washing (baptism). Further heating led to "albedo" or resurrection. The goal, redemption, was attained through "rubedo", with or without "xanthosis".

As Berthelot[2] has remarked, there was an affinity from the start between the initiation to gnosis, which enlightens the meaning of the allegories and symbols under which the teachings of philosophy and religion are hidden, and alchemy, which endeavours to unravel the hidden properties of nature and represent them by symbols understandable to the adept. But there is a difference at the start, as the method of gnostic knowledge is revelation, while the approach of the alchemist is a laboratory approach. It is what he learns in the laboratory about the mysteries of nature that the alchemist expresses in allegories and symbolism.

In Chapter 2 we have chosen, for the sake of brevity, the experimental and the symbolic aspects of the egg, but we might have elaborated at length on this relation. To take only one example, when Robert Fludd describes his experiments on wheat[3], he calls the grain of wheat "as it were, the philosophical egg of this vegetable". It should also be noted that, for the alchemists, the grain of wheat corresponds, among cereals, to gold among minerals, to the heart in the human body and to the sun among planets. As Fludd says, it is "the only Kinge of all vegetable graines beinge that in it Nature herselfe that only Queene Mother and nurse of this world is most resident" and he considers wheat as being so akin to "mans vitall and natural faculty" that it is more nourishing to man's body than is any other vegetable.

The alchemists, as recalled above, when working in the field of the mineral kingdom, began by "putrefying" the prime matter with the purpose of producing the seed of gold. What is, in this context, an allegory is taken by Fludd as a new starting point in laboratory work, and he begins his experiment by the putrefaction of wheat, not, he emphasizes, for the purpose of later resurrecting the grain, but to reduce it to principles with a view to extracting its quintessence.

It can be seen that, in the period of western European alchemy, situated between the 12th and the 16th centuries, the symbols used by alchemists had, with all due regard to the deep difference in methodology, a function analogous to the role of hypothesis in the experimental methodology of the 19th century. But later on, and after it gave birth to iatrochemistry, which would eventually replace alchemy by a true science, those who remained adepts of the old body of alchemical teachings more and more dedicated to symbolism, were less and less interested in laboratory exercises, of which a new, and much more fruitful form was developing, as chemistry, attracting the gifted men. What remained of alchemy degenerated into mere vagaries, sinking into the black mud of esoterism and of mysticism. It is with these degenerate and residual aspects that Jung's interpretations of alchemy[4] are mainly concerned. It is rather amazing to read a description of the goals of alchemy as defined by a disciple of Jung as "to reach, beyond the borders of reason's superficial lights, the obscure depth from which the new light will spout, which will reveal to man the god within himself"[5]. According to the same concepts of alchemy, the laboratory work, besides the search for metal transmutations and of macrobiotic elixirs, was intended to use up the disconcerting transformations of materials, metals or others, under the influence of fire,

"as enigmas exhausting the mind and ensuring the complete defeat of the self, which is concomitant with the hour of God and produces the spouting out, in the residual flaw between the two poles of an inexplicable contradiction, of the flash of illumination." (Perrot[5], translation by author)

In traditional alchemy, as well as the endeavour to unravel the hidden properties of nature, "transcendance" was sought, as well as a spiritual perfection of the soul, but to identify these with a quest for mystic illumination ignores the evidence. On the contrary, it may be stated that for traditional alchemists (whatever the degenerate descendants may believe) the laboratory work, far from intending to produce discouragement, moroseness and the intellectual weakness prone to mystical illumination, represented an *askesis*, a practice of austere exercises conditioning, by a contact with reality, not only the coherence of the symbolic system derived from it but the development of intellectual and spiritual faculties leading them to imagine the transformations taking place in nature. Keeping the discourse on nature in contact with nature, the practice of the alchemical workshop was an ethical method through which the ethics of knowledge is imposed on the alchemist by his own choice, as an axiomatic condition of the authenticity of any reaction or discourse on which the symbolic system of traditional alchemy is built.

It is commonly stated that the alchemists utterly failed in their attempt to isolate the *pneuma*.

References p. 330

Zum Teufel ist der Spiritus. Das Phlegma is geblieben.

This is true enough, and for reasons well known to us. But alchemy not only contributed to the practice of laboratory work, but as we are now aware, led to the conceptual developments that formed the beginnings of the science of the 16th century and accomplished the ultimate stage of the Scientific Revolution.

REFERENCES

1 H. J. Sheppard, *Ambix*, 6 (1957) 86.
2 M. Berthelot, *Les Origines de l'Alchimie*, Paris, 1885.
3 C. H. Josten, *Ambix*, 11 (1963) 1.
4 C. G. Jung, *Psychologie und Alchemie*, Zurich, 1944. 2nd ed. 1952 (English translation by R. T. C. Hull, *Psychology and Alchemy*, Princeton, 1953, 2nd ed., 1968).
5 E. Perrot, *La Voie de la Transformation d'après C. H. Jung et l'Alchimie*, Paris, 1970.

Appendix 3

Alcoholic Fermentation and the Cell Theory

The readers of Schwann's *Mikroskopische Untersuchungen* may believe that in the footnote on the experiment on yeast, pages 234–236 (cited on p. 138), Schwann is referring to experiments made at the time of the studies that led him to conceive yeast as the agent of alcoholic fermentation, and are reported in his paper of 1837. The present author will soon publish the first Notebook of Schwann on the trail of his researches on cells. In this manuscript, after relating a number of microscopical observations, an entry dated "August 1838" is found:

"*Gährung*

Um zu beweisen, dass die Zuckerpilze nicht bloss Begleiter, sondern die Ursache der Gährung sind, käme es darauf an (ausser der Verfahren mit den Giften) zu beweisen, dass die Kohlensäure und der Alkohol zunächst in den Pilzen entstehe. Mit der Kohlensäure lässt sich dies auf folgende Weise machen:

"(*1*) Man färbt eine Zuckerlösung mit Lackmus blau und bringt sehr wenig Hefe hinein und beobachtet dann unter dem Mikroskop ob die rothe Farbe zuerst um die Pilze entsteht. Oder man taucht Lackmuspapier in eine Zuckerlösung, die nur wenige Pilze enthält, faltet diese feucht und sieht, ob es sich nicht um die Pilze roth färbt. Bei dem Experimentieren mit der blau gefärbten Flüssigkeit ist es vielleicht nothwendig Gummi zuzusetzen, um das Zerstreuen der rothen Flüssigkeit durch die ganze Masse zu verhindern.

"(*2*) Man setze Kalkwasser zu der Zuckerlösung und sehe, ob die Krystalle von kohlensäurem Kalk sich zuerst um die Pilze bilden. Da man hier sagen könnte, dass das geschehe um demselben Grunde, wie sich der Krystallische Zucker an Fäden setzt, so könnte man umgekehrt, eine Zuckerlösung mit frisch nidergeschlagenem kohlensaurem Kalk vermischen und sehen ob er sich in der [...] Köhlensäure zunächst um die Pilze zugleich wieder auflöst."

This text shows the part played by the observation of yeast in the development of the theory of the cells. The metabolic developments contained in the later section are evidently based on the study of yeast cells. The experiments described in the footnote on pp. 234–236 of the *Mikroskopische Untersuchungen* were performed at the time of the preparation of Schwann's book and, taking into account the date of the text reproduced above, after August 1838.

It is in the *Mikroskopische Untersuchungen* that Schwann, differing in that respect from Cagniard-Latour, who considered fermentation as taking place in the liquid phase of the yeast suspension, located on an experimental basis the metabolic phenomena of alcoholic fermentation inside the yeast cells, a conclusion from which he derived his general concept of the intracellular location of metabolism.

Name Index

Subject Index

COMPREHENSIVE BIOCHEMISTRY

Section V — Chemical Biology

Volume 22. Bioenergetics
Volume 23. Cytochemistry
Volume 24. Biological information transfer. Immunochemistry
Volume 25. Regulatory functions, membrane phenomena
Volume 26. Part A. Extracellular and supporting structures
Volume 26. Part B. Extracellular and supporting structures (continued)
Volume 26. Part C. Extracellular and supporting structures (continued)
Volume 27. Photobiology, ionizing radiations
Volume 28. Morphogenesis, differentiation and development
Volume 29. Comparative biochemistry, molecular evolution

Section VI — A History of Biochemistry

Volume 30. *Part I.* Proto- biochemistry
Part II. From proto-biochemistry to biochemistry
Volume 31. *Part III.* History of the identification of the sources of free energy in organisms
Volume 32. *Part IV.* The unravelling of biosynthetic pathways
Volume 33. *Part V.* History of molecular interpretations of physiological and biological concepts, and of the origins of the conception of life as the expression of a molecular order

Volume 34. *General Index*